FIREWOOD

FIREWOOD

A woodcutter's fieldguide to trees
in summer and winter

M. Michaelson

and books

SOUTH BEND INDIANA

FIREWOOD

and books
702 South Michigan, South Bend, Indiana 46618

Library of Congress Catalog Number: 83-070326

International Standard Book Number: 0-89708-115-3

First printing. June, 1983.

Printed in the United States of America.

Additional copies available:

the distributors
702 South Michigan
South Bend, Indiana 46618

The author wishes to thank the Minnesota Horticultural Society for allowing use in this book of their illustrations of trees in winter.

Other identifying characteristics of trees are based on the United States Forest Service series "American Woods."

The figures appearing in charts of heat values and weights of different tree species are based primarily on data of the Forest Products Lab in Madison, Wisconsin, as reported in BTU's Bulletin 753, USDA.

Cover and additional artwork by Peter Demerest.

CONTENTS

FOREWORD 9
INTRODUCTION 11
OAKS 29
WHITE OAK 31
RED OAK 35
HICKORIES 37
SHAGBARK HICKORY 39
WHITE ASH 43
ELM 47
AMERICAN ELM 49
AMERICAN BASSWOOD 51
BLACK WALNUT 55
AMERICAN BEECH 59
BLACK LOCUST 65
BLACK WILLOW 69
COTTONWOOD 71
HACKBERRY 75
ASPEN 79
CANOE BIRCH 81
YELLOW BIRCH 83
SWEET BIRCH 87
SUGAR MAPLE 89
BOXELDER 93
BLACK CHERRY 95
FLOWERING DOGWOOD 97
BUCKEYE 101
HOP HORNBEAM 103
BUTTERNUT 105

AMERICAN HOLLY 109
TUPELO 113
YELLOW POPLAR 117
AMERICAN SYCAMORE. 121
SWEETGUM 125
OSAGE ORANGE. 127
COMMON PERSIMMON 131
PECAN 133
RED ALDER 135
MAGNOLIA. 137
TAMARACK 139
BALD CYPRESS. 143
ENGELMANN SPRUCE. 147
SITKA SPRUCE 151
BALSAM FIR. 153
INCENSE-CEDAR. 155
WESTERN WHITE PINE 159
WESTERN REDCEDAR. 163
PONDEROSA PINE. 165
ALASKA CEDAR 169
REDWOOD 171
PORT ORFORD CEDAR 175
JACK PINE 177
EASTERN WHITE PINE 179
EASTERN HEMLOCK. 183
EASTERN SPRUCE 187
WHITE SPRUCE. 189
RED SPRUCE 191
BLACK SPRUCE 193
APPENDIX 195
INDEX 197

FOREWORD

If you buy firewood, use this guide to verify that the wood you ordered is the wood actually delivered. Unscrupulous dealers have been known to peddle red elm as oak and other inferior woods as more valuable firewood.

Firewood usually is sold by volume, that is, by the cord. A cord is 4 feet high, 8 feet long, and 4 feet wide. A "face cord" is 4 feet high, 8 feet long, but only 16 inches wide, or only one third the size of a true cord. Know which of the two is being quoted when you discuss prices of firewood.

If you cut your own firewood, use this guide to distinguish between inferior and valuable species of trees. The proverb that the woodcutter is warmed twice by firewood — once when cutting it and again when burning it — is true enough regardless of the wood. But it also is true that a cord of hickory, the queen of fire wood, delivers about 25 million BTUs compared to the same amount of American elm with only 17 million BTUs and that a cord of white oak has 23 million BTUs compared to aspen's 13 million.

When given a choice between different kinds of wood, choose the firewood that delivers the most BTUs — and be warmed twice as much!

INTRODUCTION

For those who heat with wood, the North American continent is a paradise. Unlike much of the globe — Africa, India, the Middle East, and many other regions — the United States and Canada are blessed still with vast tracts of forest land.

State and national forests are open on a permit basis to the woodcutter, as are numerous small reaches of private and public land where simple courtesy can obtain admittance to a deadfall here, a dying elm there.

Oak wilt and Dutch elm disease, those plagues of the woodland, sadden the wood cutter as much as the next man; but when death comes, the oak is not left to stand like some grotesque gray skeleton in summer. It is salvaged. And the millions upon millions of dead and dying elms, though largely spurned for building material or other manufacturing uses, can provide very satisfactory firewood in seemingly limitless quantity.

Despite our wealth, the popularity of heating with wood and the rapidly escalating cost of other fuel will necessitate intelligent conservation of our forests. This, in spite of the optimistic predictions of many foresters and tree companies who contend that there is enough wood to provide fuel for hundreds of years.

Pick up any narrative of the 18th or 19th century American frontier and read the startled settlers' reactions to the "big woods." Through Kentucky, Ohio, Indiana, Michigan, Wisconsin, Minnesota, the early settler trekked mile after mile through hardwood forests so tall and thick they literally blotted out the sun.

And, as in Ohio where the vast tracts of timber were felled to prepare fields for sowing, no other solution to the surplus was conceived but that of burning in open fires, putting the torch to pyramid after pyramid of walnut, maple, oak, and hickory — so that the most vivid recollection of many frontiersmen was the constant smell of smoke in the nostrils.

The land was cleared; great prairies unravel now where impenetrable forests once stood.

In most poor countries today more than 90 percent of the people depend on fire wood as their primary fuel, but population growth is outstripping the rise of new trees.

In these countries, wood is a major living expense. In Niger, part of drought stricken West Africa, an average family spends about one quarter of its annual income on firewood. In once heavily forested Nepal, trips from the cities to gather firewood now take an entire day; one generation ago, the same foraging trips required no more than an hour or two.

In northern Africa the land around many cities is bare of trees for as much as 40 miles in every direction, and the urban population continues to spread out further and further in search of indispensable fire wood without regard to replanting of trees.

When the trees are gone, so goes the topsoil. Rainfall runs off the packed earth. When drought comes, the desert is not far behind.

Here in the United States some encourage us to rapidly harvest our timber resources. Conservation is advocated for oil and gas, not for the supposedly inexhaustible stands of timber. Following such advice will one day create a thousand dust bowls where, with wisdom, a rich and renewable resource could be maintained.

Proper conservation of trees can ensure a continuing supply of wood for millions of people, but constant vigilance will be required to prevent the deforestation besetting so much of the rest of the world.

And so, while this book is intended as a guide for the woodcutter to assist in securing the most efficient wood fuel, it also makes a plea for conservation by imparting knowledge of tree species and the characteristics of individual woods, so that the person who cooks and heats and lives with wood does so with wisdom and a sense of self-sufficiency never experienced by those who anxiously wait for the next delivery of the fuel oil truck.

WHERE TO FIND FIREWOOD

Wood can be obtained at dumps and land fills, at building projects where trees are felled to make room for construction, and from state and national forests. To cut in national forests you must apply for a permit that allows you to cut only down timber at a maximum of 10 — 20 cords per season. None may be sold.

THE BEST FIREWOODS

Softwoods like fir, spruce, and pine ignite quickly when starting a fire. They also burn fast, with a hot flame.

Hardwoods like oak, ash, and maple burn more slowly, but give off more heat when measured in BTUs per cord.

In general, the greater the weight and density of wood, the better it is as fuel. Contrary to popular belief, the names hardwood and softwood do not refer to the weight, hardness, or value of wood as fuel. In fact, such hardwood trees as cottonwood and aspen have softer wood than the white pines and true firs.

Hardwoods are those trees having broad leaves. Softwoods have needles or scale like leaves and are usually evergreen. Several softwoods actually produce more BTUs per pound than the best hardwoods.

As can be seen in the following table, equal weights of dry hardwoods give off about the same amount of heat. A pound of dry elm yields as much as a pound of white oak. But, resinous woods like pitch pine have a much greater heating value per pound because a pound of resin gives off about twice as much heat as a pound of wood.

HEATING VALUE OF WOODS

SOFTWOODS	BTUs PER POUND
Pine, pitch	11,320
Cypress	9,870
Pine, yellow	9,610
Douglas fir	9,050
Redwood	9,040
Pine, white	8,900
Western hemlock	8,620
White cedar	8,400
HARDWOODS	
Ash, white	8,920
Poplar	8,920
Elm	8,810
White oak	8,810
Beech	8,760
Red oak	8,690
Hickory	8,670
White birch	8,650
Maple	8,580

Here, though, is the crucial difference in measuring heating values of wood. Different species of trees vary enormously in weight. (Weight is based on density, or the actual amount of cell wall.) For example, a cord of hickory weighs twice as much as a cord of aspen. And . . . .most wood is bought by the cord!

A pound of one hard wood yields about the same heat as a pound of any other species of hardwood, but a log of hickory will weigh twice as much as a log of aspen and thus will deliver twice as much heat. This is the essential point to remember when considering different species of wood, as the following table reveals.

WOOD	Millions of BTUs per cord	Pounds per cord
Hickory	25	4,600
Oak, white	23	4,400
Beech	22	4,000
Maple, sugar	21	4,000
Oak, red	21	3,900
Birch, yellow	21	3,900
Pine, yellow	21	3,700
Ash, white	20	3,800
Maple, red	19	3,400
Tamarack	19	
Birch, paper	18	3,400
Fir, Douglas	18	2,900
Pine, pitch	18	3,200
Pine, Norway	18	2,800
Birch, gray	18	3,500
Elm	18	3,200
Spruce, red	15	2,600
Hemlock	15	2,600
Balsam	14	2,200
Pine, white	14	2,700
Aspen	13	2,200
Basswood	11	

Many studies on this subject have yielded varying statistics. Average figures for air dried wood are given in this chart.

Heating values of wood also are affected greatly by moisture content. Moisture has to be driven off before a fire can begin to generate significant heat. When a fire reaches 400 degrees, gases begin to be driven off. Then, at about 600 degrees, the wood itself burns. When finally the fire reaches 1400 degrees gases begin to ignite and full combustion takes effect. This is the point at which a truly hot fire is born, and this is the reason it is so important to burn only well-dried wood that has been seasoned over a spring and summer, or even better, over an entire year. A cord of well seasoned wood like hickory is equal in heating value to one ton of coal.

SPLITTING, STACKING & DRYING WOOD

Live trees have considerable moisture content and to prepare them for suitable firewood they must be split and dried before burning. Some woods, like aspen, are easy to split; some, like American elm, are solid cement.

Keep two tools within reach when splitting wood. First, have a very sharp, fat heeled axe for the easy-to-split woods. For larger and more difficult logs, use a maul or a sledge hammer and wedge.

Anyone who for the first time watches an experienced logger split wood is amazed at the ease and economy of effort with which even large logs can be split. The novice rears back with the axe and strikes the log a terrific blow that rebounds from neck to toe. He expends twice the energy required to do the job, and the next day his back feels as if it has been tortured on a rack.

Instead of striking each log with maximum effort, try to size up the piece of wood. Where is the grain? Are there cracks that will split more easily, or knots to prevent splits? Then swing evenly with the axe or maul. You can adjust your efforts gradually upward until you reach the most effective force needed to split specific types of wood. You will be amazed at how much more wood you can split with minimum effort by following this method.

The ease or difficulty of splitting various types of wood depends upon the grain of the wood, the quantity of knots, and the temperature.

Wood having straight grain is the most easily split. Wood having interlocked grain is the most difficult to split. This is true of gums, cottonwoods, mahogany, beech, sycamores, and elms. Also, seasoned wood is usually more difficult to split than green logs. Knots obstruct splitting and cause spitting when burned. And frozen wood can be split much more easily than unfrozen wood.

With the more difficult-to-split logs it may help to notch logs with a chainsaw and then drive a wedge into the opening made by the saw. Chainsaws are not designed to rip through the middle of a log, but if that is the only way to get the job done, do it — but carefully!

After splitting the wood, it should be stacked so that air circulates around all surfaces for fast, efficient drying. If you place a number of logs on the ground and build a stack of split wood on top by continuing to place two pieces at right angles as high as you can reach, you will have a very loose stack but one that will season very quickly. Another, slightly less efficient system but one that takes up less space, is to place several pieces on each level without leaving as much leeway between logs.

Green wood may be about 50 percent moisture by weight and will weigh more than air dried wood which should average 15-25 percent moisture. Moisture in fire wood greatly diminishes its value as fuel. Absolutely dry wood could have 8,000 BTUs per pound, but 25 percent moisture wood has about 6,000 BTUs per pound.

When wood is sufficiently dry to burn, the wood will darken in color and large cracks often will be visible at the butt ends of logs.

STARTING A FIRE

Few green trees can be burned satisfactorily. They sputter and stall and, when ignited, give off too much smoke and creosote that gums up the chimney. Use dry well seasoned wood!

Roll newspapers into balls. Place one large log in back of the fireplace and one in front. Put newspapers between these logs with kindling on top.

To create an effective draft and get the fire off to a fast start, take a section of loose newspaper and place it on top of the logs. When ignited, this loose piece will burn very rapidly and create a strong draft up the chimney.

The remaining newspapers and kindling will burn more rapidly with a strong draft. Keep the damper wide open. Fire can not live without oxygen. If possible, open a window in the room where a fire is started until the logs are blazing. If the house is caulked and weatherstripped and airtight, you may have to occasionally reopen a window to keep the fire burning vigorously.

The ideal situation is to have a fireplace in a room that can be closed off, since to live the fire must draw oxygen from throughout the house, and if the fireplace is in an open location it will draw off heated air from the remainder of the house including the upstairs bedrooms.

Also, it is wise to close off the fireplace room when the fire is dying out, since warm air will continue to rise through the chimney during the night if the damper is left open. This is one more argument in favor of wood burning stoves or furnaces versus fireplaces. In general, the stoves are far more efficient and use much less wood than do fireplaces.

Whatever you use to heat with wood, be convinced that power blackouts, blizzards, and crippling oil and gas prices no longer hold the kind of misery for you that others experience. For to live with a natural, self-renewable resource like wood is to return to a more practical self sufficiency, yielding a confidence and satisfaction that words can not adequately describe—that only experience can properly convey.

IDENTIFYING TREES IN SUMMER AND WINTER

Evergreens can be easily identified by comparing needles. Similarly, deciduous trees are identified by their leaves.

Needles may stay on a tree for many years in succession, while leaves on deciduous trees fall and are renewed during the following growing season.

There are two basic types of leaves; (1) The single leaf as in oaks, which is called simple, and (2) the leaves of hickories and walnuts that are called compound and are made up of several different leaflets.

(1) Simple

(2) Compound

Another key to identification is the shape of the leaf. The most common forms are lanceolate, oblong, elliptical, and oval.

Lanceolate

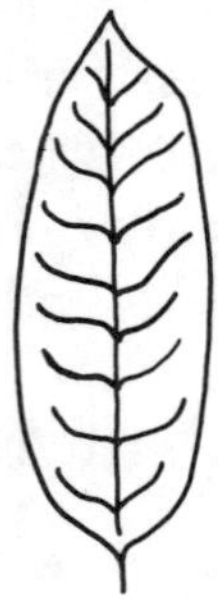

Oblong

Elliptical

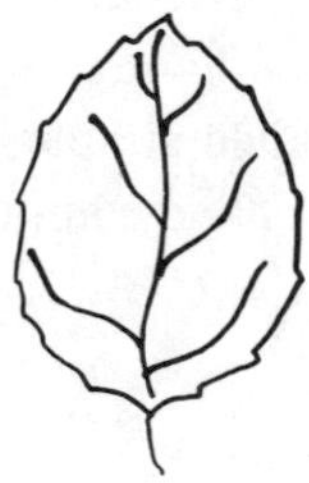

Oval

The third key to leaves is whether they are opposite or alternate. If two leaves are directly opposite on the stem leaf, arrangement is said to be opposite. If, however, leaves are staggered, leaf arrangement is called alternate.

Other keys to identification are present many in summer, fewer in winter. The silhouette of the tree, its size and shape, the relationship of the trunk and crown, spread of branches (horizontal, upright, or gnarled), an open or densely twigged crown, bud color and pattern, presence of thorns, fruit, leaf scars, flowers — all can be used to identify valuable species for firewood.

Leaves, fruit, and flowers are commonly used to identify trees in summer. But for the woodcutter, who usually works in winter or spring before leaves have sprouted and who is not primarily studying nature but determined upon a more utilitarian endeavor, identification usually depends on sizing up the shape of a tree and its bark.

Another valuable source of identification is the scar left on stems by leaves as they drop off in autumn. These are called leaf scars. In different species they may be round, triangular, narrow, oval, horseshoe shaped, or heart shaped. They can lie flat on the stem, or be concave or convex. and like leaves, leaf scars may be opposite on the stems as in maples or ashes, or alternate as in hickories, walnuts, and oaks.

Comparisons of the sapwood and heartwood in logs also is a key to identification. Logs of many species have an outer light colored section (sapwood) and a central, darker core (heartwood).

Cells in the sapwood carry on the life of a tree by storing food and conducting sap. Only the cells storing food are alive. The cells in heartwood are all dead and are valuable only in providing strength to the tree.

In a continuous process, inner sapwood changes gradually to heartwood and becomes filled with gums, resins, and other coloring matter. These substances make the wood darker and increase resistance to decay.

Among some trees there is scant difference in color between heartwood and sapwood. This is true of beech, cottonwood, tupelo, basswood, hackberry, holly, aspen, fir, balsam, hemlock, and spruce. But in hickory, white ash, maple, dogwood, and persimmon, the sapwood is so wide it comprises the major part of lumber cut from those trees.

So, in winter the woodcutter should begin by first sizing up the shape of the tree and then continue by examining the bark.

Tree bark differs widely among various species and even among young and old trees of the same species. But among older trees there usually is some significant characteristic that can be used by the experienced woodsman to identify particular species.

On some trees like the beech, bark will be hard and smooth. On other trees, bark can be flaked off in rough plates like the shagbark hickory, or in thin brittle strips like the hornbeam, or in still others peeled off laterally like a canoe birch. Some of the easiest trees to recognize by bark alone are the sycamore, with its distinctive mottled, flaky appearance and the shagbark hickory with bark hanging in long, loose, shaggy plates.

Although the bark coloration of trees of the same species may vary among young and old trees and even among trees growing in different locations, color can still play an important role when combined with other keys to identification.

The cracks called fissures or furrows bordered by ridges that develop as the bark grows also help to tell different species apart. These furrows and ridges form different patterns, like the distinctive interlacing diamond shaped ridges of the white ash, or the peculiar blocks called alligator bark on the persimmon and dogwood.

A chart is provided in the following pages to help key the color and texture of bark according to different species.

When bark characteristics are added to the photographs of trees in winter, comparisons of different trees readily take shape in the woodcutters eye and oak is soon distinguished from elm, ash from cottonwood, and hickory from black willow.

After starting to practice using these keys to identification, the time will steadily grow shorter between sizing up a tree for identification and the subsequent roar as the chainsaw swings into action.

One of the best characteristics for identification of trees in winter is bark color and texture. On this and on the continuing two pages are illustrations of different types of bark, followed by a chart of keys to be used when distinguishing different trees in the field.

Words like ridged, fissured, and furrowed often are used very carelessly and even interchangeably in some books.

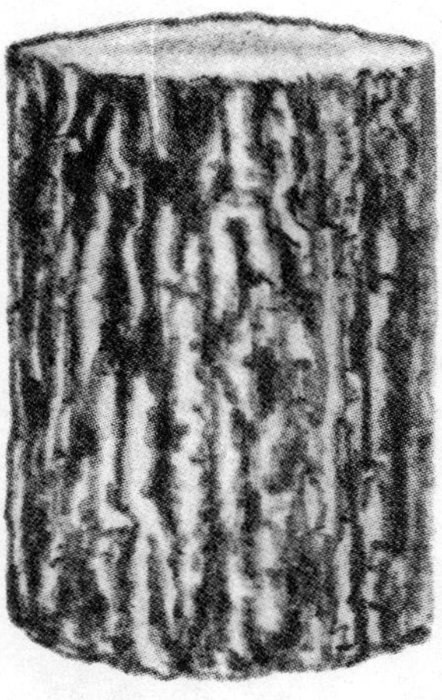

FURROWED: in this guide, is used to describe bark having cracks (fissures) or deep grooves separating elevated ridges.

SMOOTH: like the beech, is bark that has no furrows or other identifying marks.

WARTY: is bark that resembles a toad's back, with small irregular, dark growths like the hackberry.

SHAGGY: is bark that presents a very rough or shaggy appearance like the shagbark hickory.

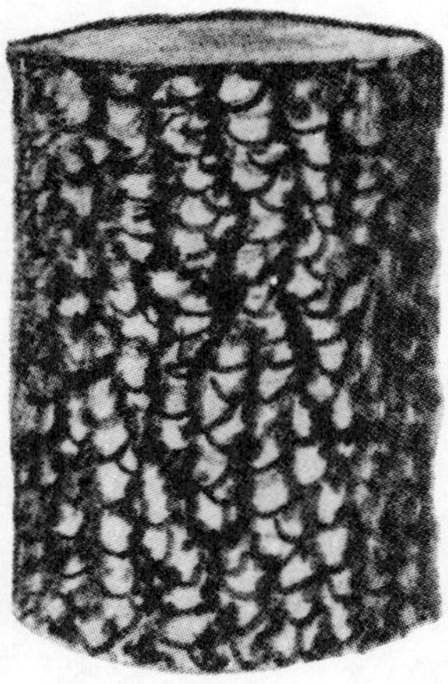

QUADRANGULAR: blocks, or alligator bark, is divided into squares like the persimmon.

The descriptions of color used in the following chart are the most common color of the species as mature trees. As mentioned earlier, color often varies in the same species. But there are typical bark colors for most species and the most typical and most commonly used descriptions are listed in the chart. Use it in conjunction with the photographs for more accurate identification.

BARK CHART

COLOR	FURROWED	SMOOTH	SCALY (Flaky)	SHAGGY	WARTY	QUADRANGULAR BLOCKS
GRAY	YELLOW POPLAR (interlacing ridges) BUTTERNUT ASH (interlocking diamond shaped ridges) ELM (also scaly)	BEECH RED ALDER	SUGAR MAPLE ELM	HICKORY	HOLLY	
DARK GRAY	COTTONWOOD (deep grooves) BASSWOOD	HORNBEAM	MAGNOLIA			
BROWN	RED OAK		HOP HORNBEAM (Ironwood) PECAN (Interlocked ridges)			TUPELO
GRAYISH BROWN	SWEETGUM BUCKEYE (scaly) WHITE OAK (scaly)		BUCKEYE WHITE OAK		HACKBERRY	
YELLOWISH BROWN		QUAKING ASPEN				
ORANGE BROWN	OSAGE ORANGE					
BLACK	BLACK LOCUST (forked ridges) BLACK WALNUT (scaly) BLACK WILLOW		BLACK WALNUT			PERSIMMON DOGWOOD
REDDISH BLACK			CHERRY			
WHITISH GREEN	ASPEN					
WHITE	BIRCH					

OAKS

There are over fifty species of native oaks in the United States, but when using oak for firewood it is necessary to distinguish between only two main groups—red oaks and white oaks.

Black Oak, grown in the open has a widespreading crown of large ascending branches.

The red oak group includes northern red oak, pin oak, black oak, turkey oak, live oak, southern red oak, swamp red oak, water oak, blackjack oak, willow oak and laurel oak.

The white oak group encompasses true white oak, post oak, bur oak, swamp white oak, swamp chestnut oak, chestnut oak, overcup oak, and Oregon white oak.

The chief characteristics separating the red oak group from the white oak are color of bark, color of interior bark and pores which appear as tiny holes on a smoothly cut end surface of a log.

Red oak has dark brown to black bark while white oak has a grayish color. The heartwood of red oak has a reddish tinge, while white oak is grayish brown. And the large pores in the heartwood and in inner sapwood of the white oak group are usually plugged with a froth-like growth called tyloses, while the pores of the red oak group are open.

When given a choice, white oak should be selected over red for firewood. The former yields 23 million BTUs per cord, compared to about 21 million for red oak. However, red oak also is a prized firewood. It is easy to split and outperforms many other woods in the fireplace.

WHITE OAK

SHAPE: One of the most valuable and beautiful of all species. White oaks may reach heights of 150 feet with a trunk 4-5 feet in diameter. Given room to grow, it will form into a broad, full top.

BARK: Light, grayish brown in color. Scaly and furrowed.

LEAVES: Dark green on top, paler underside, 5-9 inches long and 3-6 inches wide, with 5 to 9 rounded lobes. Alternate.

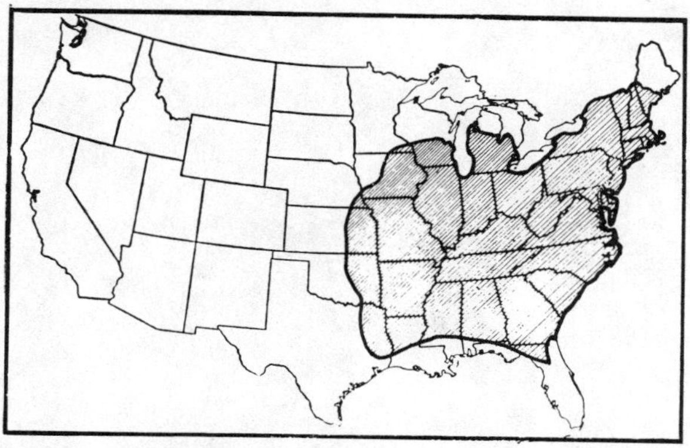

White oak probably got its name from its light gray bark and not from the color of its heartwood, which varies from a rich, light or grayish brown, to dark brown. The red tinge characteristic of red oak heartwood is usually absent in the white oaks and white oak growth rings are very distinct.

White Oak, grown in the open, displays a low crown which increases in breadth with age.

White oak is very strong, dense, and heavy. It is one of the best overall firewoods.

White Oak bark is characterized by its light ash-gray color and scaly plates.

White Oak leaves are glossy with five to nine rounded lobes. One quarter of the nut is covered by a shallow cap

Red Oak, grown in the open, shows a relatively short trunk dividing into several sturdy branches.

RED OAK

SHAPE: rounded top with heights to 100 feet. Trunk is 1-3 feet in diameter.

BARK: brown to black, ridged and furrowed.

LEAVES: Leaves dark green above, paler below, 4-8 inches long and 4-5 inches wide. Alternate.

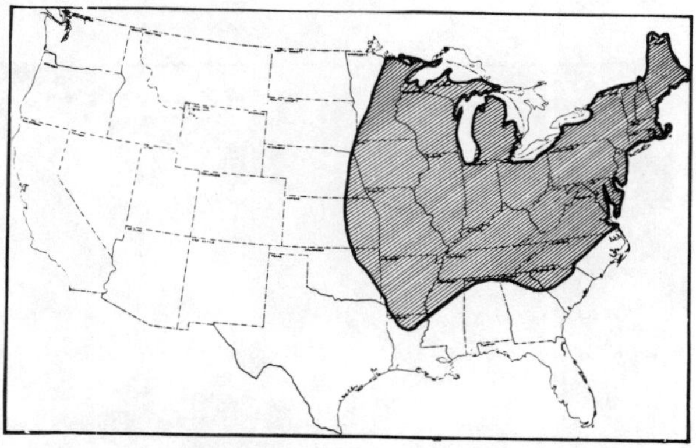

Distinguish red from white oaks by the color of bark and presence or absence of pores. Red oak has dark brown to black bark while white oak is grayish and scaly and the heartwood of red oak has a reddish tinge while white oak is grayish brown.

Red Oak bark.

Red Oak leaves, twig and acorns.

HICKORIES

A tree in the true hickory group is a medium sized deciduous hardwood. It is generally slow growing, but does respond favorably to increased soil fertility and has a reputation for long life. Some trees have been known to live for 350 years.

There are in the United States four species that are known as true hickories; shellbark, pignut, shagbark, and mockernut.

The geographic range of true hickories extends from central Maine south through the eastern United States to Mexico. Hickories grow in every state east of the Mississippi and in a few states west of it.

Shellbark hickory is not as widely distributed as the other hickories nor is it as common. The major concentrations are along the Ohio and Mississippi River valleys from Ohio and Indiana to northeastern Arkansas. Shellbark looks much like shagbark hickory except that the leaves of the former are usually 15-22 inches long in 5-9 leaflets.

Pignut is similar to shagbark in distribution; however, unlike shagbark, pignut hickory extends into the coastal plains of the southeast. On larger pignut trees the bark is broken by scaly ridges that form a rough, diamond shaped pattern. Leaves are 6-12 inches long, in 5-7 leaflets.

Mockernut hickory ranges from New Hampshire west to Iowa, south to Texas and east to Florida. Leaves are 8-12 inches long with 7-9 leaflets.

Shagbark is the most widely distributed of the true hickories. Its leaves are about 8-15 inches long with 5-9 leaflets. Young trees have smooth gray bark; older trees have shaggy bark.

The fruit is a round or egg shaped nut 1-2 inches in diameter. It is covered by a hard husk which splits lengthwise along the seams of the husk. The meaty part of the nut is edible although sometimes it is bitter. The size of the seeds or nuts varies considerably between species: shellbark has the largest, pignut has the smallest, mockernut has about 90 seeds per pound and shagbark averages 100 seeds per pound.

The bark of young hickories is usually smooth, gray, and very hard. As the tree grows older the bark either breaks up into plates or forms ridges that are separated by narrow fissures. Shagbark develops plates, wide and curled at the ends in a shaggy pattern. Shellbark is similar to shagbark except the plates are thinner and more firm. Mockernut and pignut have furrowed bark with interlacing ridges. Mockernut has firm, almost horny ridges, but pignut has somewhat scaly ridges.

HICKORY (SHAGBARK)

SHAPE: Hickories may reach 100 feet in height and have a trunk 1-2 feet in diameter. The crown is rather narrow except when the tree grows in the open.

BARK: The bark of young shagbark hickories is smooth and gray; in older trees it is distinctly shaggy with thin narrow scales curved outward at the ends.

LEAVES: Compound leaves in 5-9 leaflets, 8-15 inches long. Opposite.

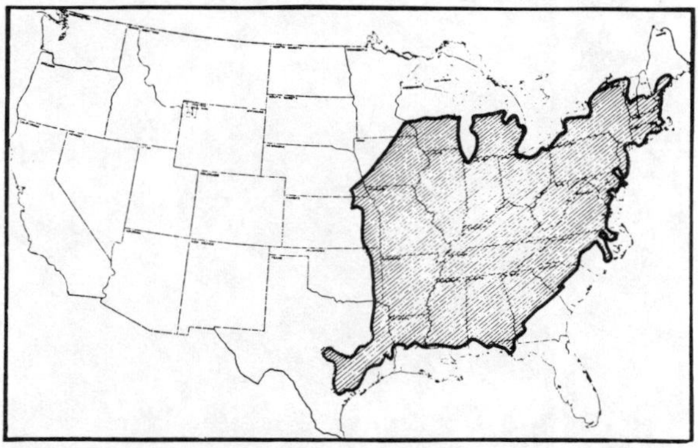

The growth rings of hickory are very noticeable, as are the heartwood and the sapwood. The heartwood is pale brown to reddish brown; the sapwood is white.

The wood of true hickories is famous for strength. Some

Shagbark Hickory has a strong trunk twenty to thirty inches in diameter. It can reach over 100 feet in height.

woods are stronger than hickory and other are harder, but no other commercial species of wood is equal to it in combined strength, toughness, hardness, and stiffness.

Hickory is the queen of firewoods, yielding 25 million BTUs per cord.

Shagbark Hickory bark with its long, loose plates characteristic of the mature tree.

Shagbark Hickory has a compound leaf generally comprised of five leaflets although it can contain up to nine.

Shagbark Hickory nut enclosed in its husk.

ASH (WHITE)

SHAPE: Grows to 120 feet with a tall, thick trunk 5-6 feet in diameter. In a forest, the crown is narrow; growing in the open, the crown is pyramidal or round.

BARK: Ash bark is usually gray. On older trees it is finely furrowed into diamond shaped, narrow, interlacing ridges.

LEAVES: All of the ashes have compound leaves. White ash leaves are 8-12 inches long with 5-9 (usually 7) leaflets, 1-3 inches wide. Opposite.

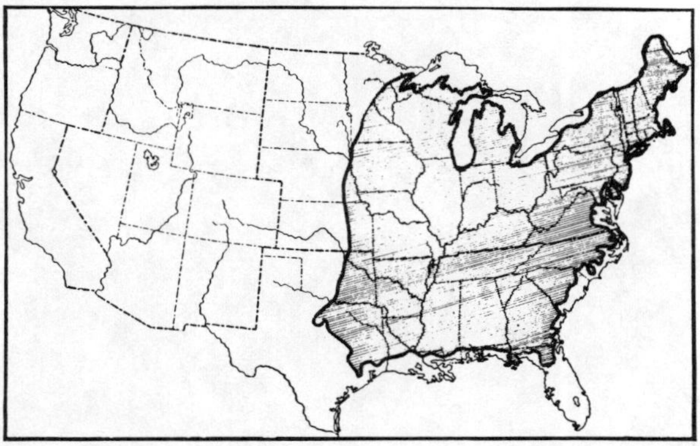

White Ash shows a compact oval crown when grown in the open.. The forest grown tree usually has a narrower crown.

The trees and shrubs of the ash family consist of about 65 species, 16 of which are found in the United States. Only six — white, pumpkin, blue, black, green, and Oregon ash — are used for lumber and other wood products.

All of the ash species have opposite, compound leaves.

The seeds are found in samaras that are clustered. The major differences helpful to identifying species are number of leaflets, whether the margins of the leaflets are smooth or toothed, and the extent to which the wing surrounds the seed.

The heartwood of ash varies from brown to grayish brown and contrasts sharply with the light colored sapwood. The growth rings are distinct.

The wood of white ash is usually straight grained, very heavy, hard and strong. Oregon ash is closest to white ash in strength; the other species are somewhat lighter and less valuable.

White ash is an excellent firewood, yielding about 20 million BTUs per cord.

White Ash bark with deep diamond shaped fissures.

White Ash leaves have opposite compound leaflets. The leaves and seeds are typical of members of the ash family.

ELM

Of the six species of elms native to North America — American, slippery, rock, winged, cedar, and September — American elm is the best known. American elm also is known as soft elm, water elm, and white elm. Rock elm may be called gray elm, or red or soft elm. Other names for winged elm are cork elm and wahoo. Red elm also is used for September elm and for cedar elm.

American Elm in winter displays divided trunk with gracefully drooping limbs. The broad crown gives the tree a vase-like appearance.

ELM (AMERICAN)

SHAPE: Trunk divides into many large, drooping limbs which develop into a graceful, vase shaped, broad topped crown. Height reaches 120 feet with a trunk 2-4 feet in diameter.

BARK: Dark gray, rough, flaky and ridged.

LEAVES: Oval, jagged, abruptly pointed leaves, dark green and 4-6 inches long, 1-3 wide. Alternate.

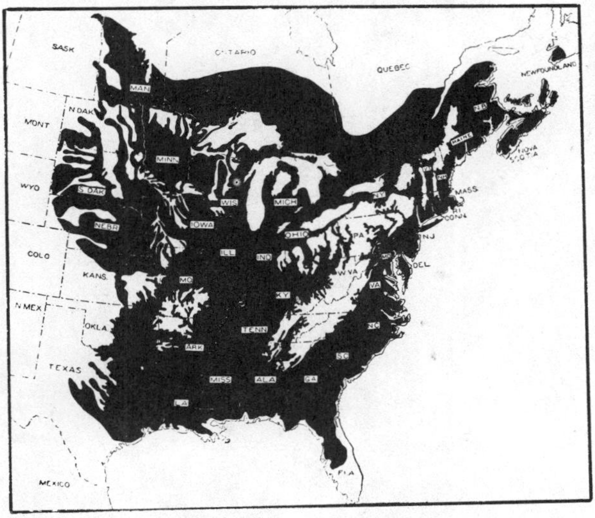

The heartwood of elm is light brown to brown, often having a reddish tinge, while the sapwood is nearly white. The annual rings in American elm and slippery elm are wide and distinct.

American elm and rock elm are very difficult to split because of their interlocked grain structure. However, the elms are fairly high in BTU yield, and because of the enormous quantity available at low cost in many regions of the United States, elms will be utilized extensively for fuel for many years to come.

American Elm bark is ashy gray with flaky ridges.

American Elm leaves are dark green with a sawtooth edge.

AMERICAN BASSWOOD

SHAPE: Basswood can reach a height of 120 feet with trunk diameters of 48-54 inches. In forests the trees usually have straight limb-free trunks and narrow, short crowns.

BARK: On young trees, the bark is grayish green and shiny. On older trees, it becomes dark gray with deep long furrows and prominent ridges.

LEAVES: 5-6 inches long and 3-4 inches wide, ovate leaves in coarsely toothed pattern. Alternate.

COMMON NAMES: American linden, linden, basswood.

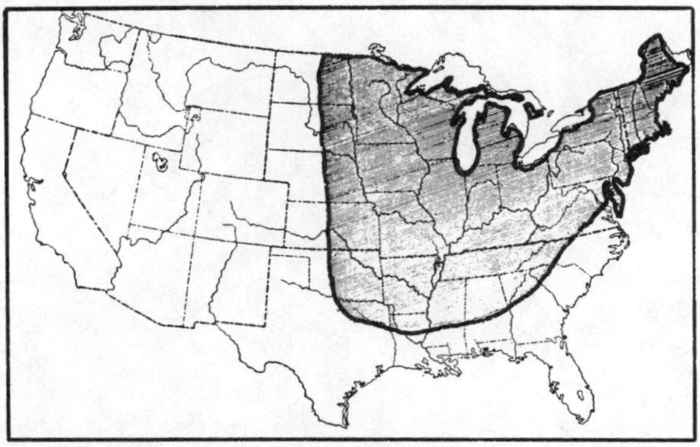

The natural range of American basswood extends from southwestern New Brunswick through central Quebec to Manitoba then south to Oklahoma. The southern limit is Arkansas and Tennessee.

American Basswood with a compact, narrow crown and straight trunk.

American basswood grows as a principal species in only two forest types: the sugar maple-basswood type in western Wisconsin and central Minnesota, and the northern red oak-basswood-white ash type common to parts of New England, Pennsylvania and the southern Appalachians.

The sapwood of basswood is whitish to creamy white and merges gradually into the pale brown or reddish heartwood which may have dark streaks.

Basswood is soft and light and weak and is not a preferred firewood, although it does burn moderately well when green compared to many other woods.

American Basswood bark is thick and furrowed.

American Basswood leaves are broad and heart-shaped. Clusters of pea-sized fruit hang from a leafy bracht.

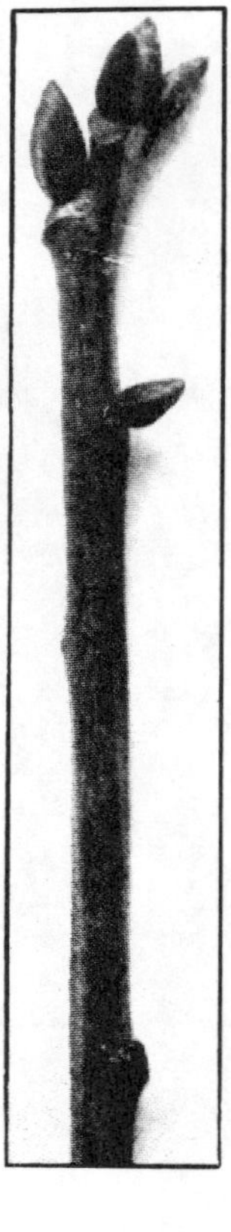

American Basswood twig bearing lateral buds arranged alternately on the twig.

BLACK WALNUT

SHAPE: Mature trees may grow to 120 feet and 30-40 inches in diameter, although some exceptional trees have been measured at 150 feet and 100 inches. In forests black walnut stems grow straight and clean; trees growing in the open often are forked and many limbed.

BARK: On older trees, bark is furrowed and dark brown to black.

LEAVES: Leaves are compound 12-24 inches long and composed of 15-23 pointed, saw-toothed leaflets paired except at the end. Alternate.

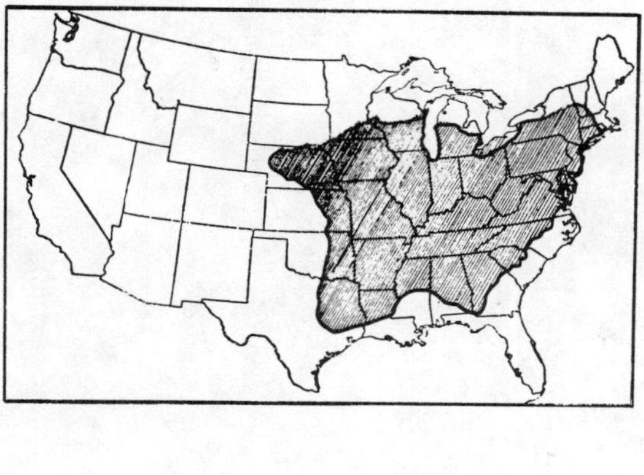

Black walnut is found throughout the central and eastern sections of the U.S. Pure stands of walnut are rare, small, and

Black Walnut tree with relatively sparse foliage and sturdy branches.

occur as groves at the edge of hardwood forests. It typically grows as scattered individual trees or in small groups in mixture with a great variety of other hardwoods.

The flowers of black walnut appear in April or May. The male flowers develop as 2-4 inch long catkins on one-year-old twigs, while one or two small, green female flowers develop on the terminals of new twig growth. The fruit, 1-3 inches in diameter, matures early in autumn and has a fleshy husk, hard shell, and an oily, edible seed.

Black walnut is one of the more rapidly growing hardwoods. On good sites young trees may grow 3 or 4 feet per year, and in 20 years may attain heights of 40-50 feet with diameters of 6-10 inches.

Black walnut is a heavy wood averaging 38 pounds per cubic foot and very hard, strong, and durable. The heartwood is chocolate brown and occasionally has darker sometimes purplish streaks. The sapwood is nearly white.

Black walnut is seldom used for fuel because of its enormously high value for quality furniture, cabinets, and gun stocks. However, it also is an excellent firewood when used for that purpose.

Black Walnut leaves are compound with paired leaflets. Two maturing walnuts are shown in their green husks.

Black Walnut nut has a hard and deeply chisled shell.

AMERICAN BEECH

SHAPE: Mature beech trees are usually 80-120 feet tall and 15-50 inches in diameter, having short trunks and broad, rounded crowns.

BARK: Thin, smooth, light, bluish gray.

LEAVES: 2-6 inches long, 1-3 inches wide; elliptical to oblong-ovate in shape with sharp teeth along the margins. Shiny surface above and below. Alternate.

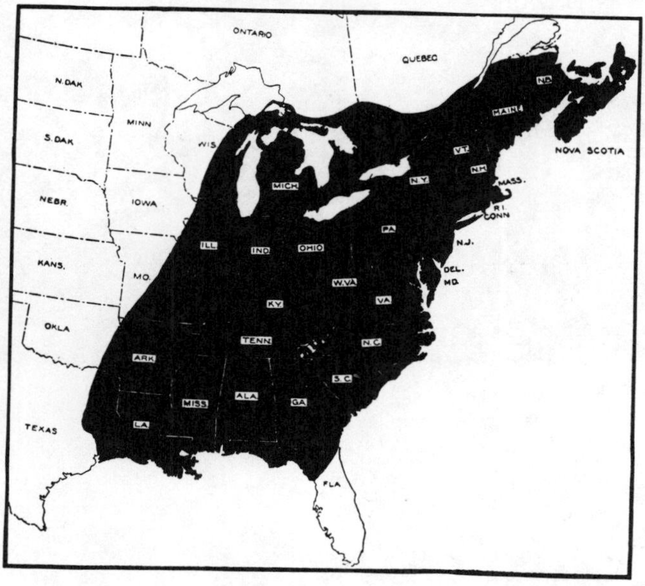

American Beech has a low, wide, spreading crown with many horizontal branches.

Beech grows in the southeastern part of Canada and the eastern half of the United States. In forests they often have narrow, compact crowns with long, clean, straight trunks. Trees growing in the open develop short, thick trunks with large, low, spreading limbs terminating in slender, somewhat drooping branches and forming broad round topped heads.

At about 40 years of age, beech trees begin to produce seeds in the form of small nuts in egg shaped prickly husks. This edible nut is thick shelled and chestnut brown.

Beech is one of the heaviest native hardwoods, with an average weight of 45 pounds per cubic foot. It is very hard and strong. The heartwood is reddish brown to whitish; the sapwood is nearly white, tinged with red.

Though difficult to split, beech is an outstanding firewood averaging 22 million BTUs per cord, excelled as a fuel only by hickory and white oak.

American Beech bark is thin and smooth with irregular dark blotches.

American Beech leaves with yellow-green blooms which appear while leaves are still limp (above). The tapered bud is an inch long and shiny brown (right).

Above: Winter photo of open-grown Beech.

Right: American Beech nuts are triangular and encased in a prickly bur.

Black Locust grown in the open has an uneven crown with a plume-like appearance.

BLACK LOCUST

SHAPE: Grows to a height of 40-100 feet with a diameter of 1-3 feet. The trunk is usually straight in the forest, but in the open it is likely to fork and be limby.

BARK: On mature trees bark is dark reddish brown to black, thick, and deeply furrowed into forked ridges.

LEAVES: Compound leaves 8-14 inches long that end in sharp points. Alternate.

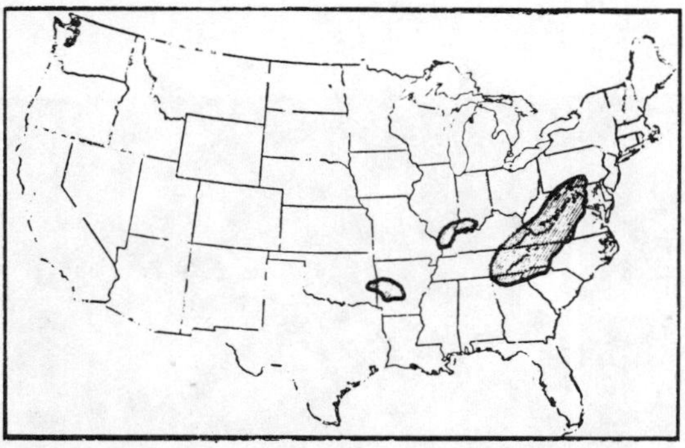

Black locust is native to the Appalachian Mountains. It is also found in many midwestern states and other regions where it has been introduced.

Black locust is distinguished by the paired, sharp spines about one-half inch long that develop at the base of each leaf. White to pink flower clusters, 4-5 inches, long appear in the spring and have a heavy, pleasant fragrance. The flat brown seed pods, one-half inch wide and 2-4 inches long, mature in the fall.

The outstanding characteristic of black locust is the durability of its heartwood. It varies in color from greenish yellow to dark brown when freshly cut. It turns to a dark brown after exposure to air. The narrow sapwood is a creamy white.

Black locust wood is quite strong, heavy, and dense. It is an excellent firewood.

Black Locust bark is dark red to black with deep, thick furrows.

Black locust leaves are compound with alternate placement. Each twig bears seven to nineteen oval leaflets.

Black Locust seed pods with four to eight seeds.

BLACK WILLOW

SHAPE: Largest of American willows, capable of attaining a height of more than 100 feet.

BARK: Rough dark brown or almost black and prominently ridged.

LEAVES: Lance shaped leaves 3-6 inches long, one-half to one inch wide, smooth and bright green above and paler underneath. Alternate.

Black willow is found throughout the eastern half of the United States. The northern limit of its range closely follows the 45th parallel from central Minnesota and Wisconsin east to the coast of Maine.

Throughout its range, black willow is a common tree in wet areas that are not permanently flooded. It grows near rivers and lakebeds where the soil is not very sandy. It also is common in swamps, sloughs, and on banks of gullies and drainage ditches.

The heartwood of black willow is pale reddish brown. The wood is moderately soft.

Black Willow bark with prominent ridges.

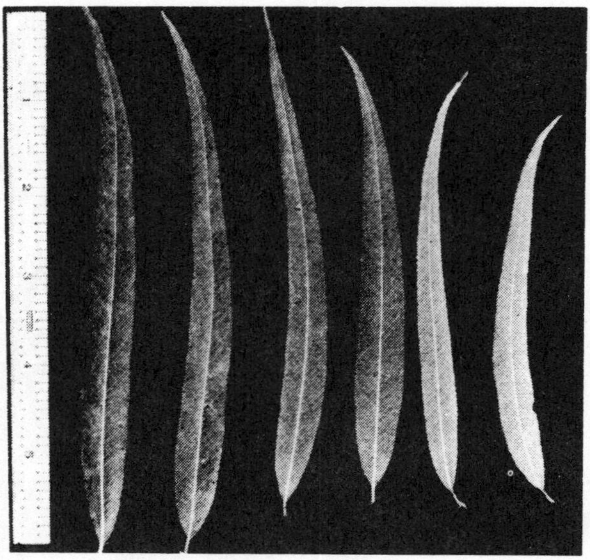

Black Willow leaves are long and sword-shaped.

COTTONWOOD

SHAPE: Largest of the poplars, to heights of 200 feet. Trunk usually 3-5 feet in diameter.

BARK: Bark of older trees is pale or dark gray and furrowed. Upper branches and young trees usually are smooth and greenish yellow.

LEAVES: Broadly triangular leaves, green above and pale underneath, 3-6 inches long and 4-5 inches wide. Alternate.

COMMON NAMES: Eastern poplar, Carolina poplar, necklace poplar, California poplar, and western balsam poplar.

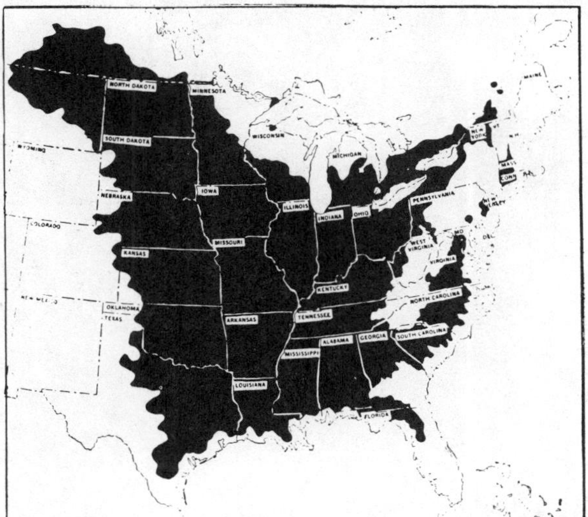

Cottonwood generally grows in the open prairies and plains.

Cottonwood is the general name used for about a dozen closely related species of the rapidly growing poplar family native to the United States.

Eastern cottonwood and black cottonwood, a western species, have the most commercial importance. Eastern cottonwood is the fastest growing tree in North America, averaging 20 inches in diameter at the trunk and 100-200 feet in height at 35 years of age.

Cottonwood is often found in bottom lands near rivers and other sources of water. Its wood is used extensively for pulpwood. It is difficult to split, soft, light, and vulnerable to decay, and therefore is one of the poorest species of firewood.

Cottonwood bark is gray with deep furrows.

Cottonwood leaves are broad and triangular, dark green above and pale underneath.

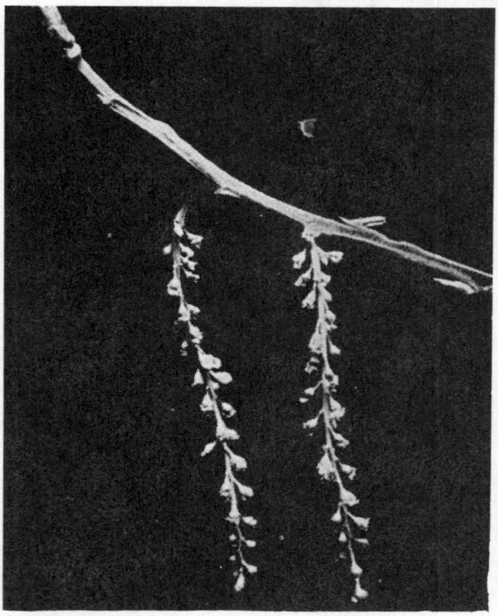

Cottonwood capsule fruits grow in long strands.

HACKBERRY (SUGARBERRY)

SHAPE: Spreading and somewhat drooping branches, with a narrow, round topped crown. Mature trees usually reach 50 feet, sometimes higher,with a trunk 1-2 feet in diameter.

BARK: Single, most prominent feature for identification is the warty, grayish brown bark.

LEAVES: Ovate, toothed leaves 2-4 inches long and 1-2 inches wide, smooth and dark green above and pale below. Alternate.

COMMON NAMES: Sugarberry, nettle tree, and false elm.

Hackberry is a name commonly applied to two closely related species —– hackberry and sugarberry. It is found from eastern North Dakota south to central Oklahoma and east to Tennessee, western North Carolina and southern New England.

Hackberry grows on a wide variety of soils, but does best on rich, moist ground. It is frequently found on slopes and bluffs and on soils with a high lime content. It is drought resistent and is used extensively in windbreaks in the Midwest.

Hackberry is a straight grained, sometimes interlocked grained, moderately hard wood with a large area of sapwood varying in color from pale yellow and often discolored with blue sap stain. Growth rings are verv prominent.

Hackberry with its drooping leaves and round crown can be found in central United States.

Hackberry is a good, fairly hard, moderately easy to split firewood.

Hackberry bark is often warty and grayish brown.

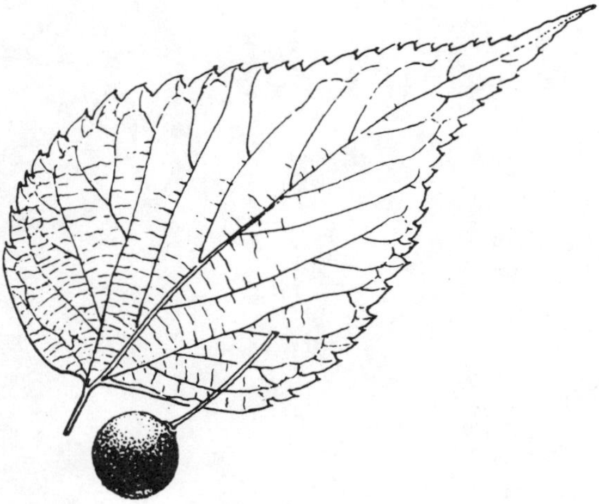

Hackberry leaves are ovate and dark green above and pale green below.

ASPEN

SHAPE: Trunks usually are straight and clear of limbs over much of their length. Grows from 20-60 feet tall and 1-2 feet in diameter.

LEAVES: Vary from broadly ovate to nearly round, with small rounded teeth. Alternate.

BARK: Varies from chalky white to green and becomes furrowed on the lower trunks of large trees. In the west the bark is often yellow-brown and furrowed.

COMMON NAMES: Quaking aspen, popple, quaky. Bigtooth is called largetooth aspen or popple.

Aspen is the most widely distributed tree species in North America. There are two main species —— quaking and bigtooth aspen. Quaking aspen ranges from Alaska through Canada into the northeastern and western United States. Bigtooth ranges from southern Canada and the northern United States, and from the Atlantic coast west to the prairies.

The woods of quaking and bigtooth aspen are similar in appearance and properties. The heartwood is nearly white to light brown. The lighter colored sapwood generally blends into the heartwood.

Aspen is generally straight grained and is one of the softer and lighter hardwoods.

Aspen bark is furrowed on the lower trunks of large trees.

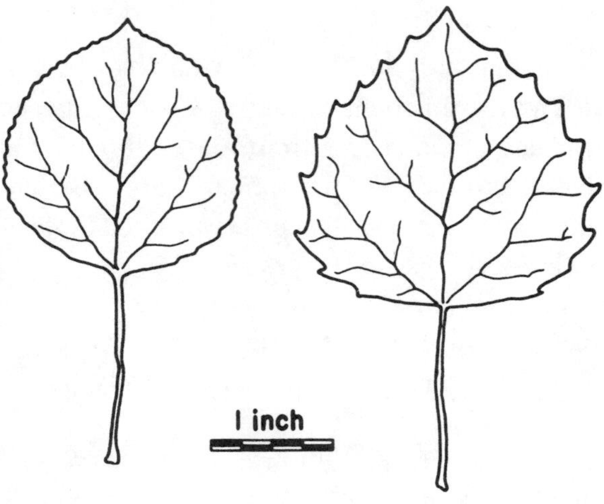

Leaves of Quaking Aspen (left) and Bigtooth Aspen (right).

CANOE BIRCH (PAPER BIRCH)

SHAPE: Graceful tree with slightly hanging branches and irregular, rounded top. Sometimes reaches height of 100 feet.

BARK: White, peeling into narrow, long strips that curl at the ends. The bark of older trees is black and furrowed at the base.

LEAVES: Dull, dark green on top; yellow green below. 2-4 inches long and 1-2 inches wide. Alternate.

Throughout the arctic regions and cooler parts of the North Temperate Zone there are approximately 40 species of trees and shrubs included in the birch group. Paper birch is the most widely distributed of the birches. It is most easily recognized by its creamy white bark that separates readily into thin papery layers.

Above: Paper Birch can be recognized by its creamy white bark that separates readily into thin papery layers.

Right: Paper Birch leaves are green on top and yellow-green underneath.

YELLOW BIRCH

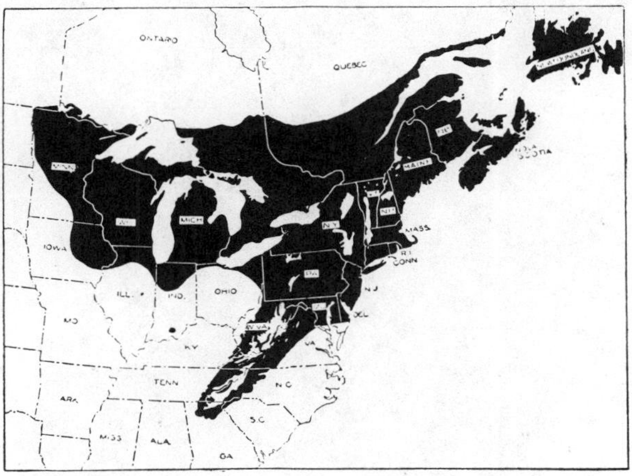

Yellow birch has leaves similar to paper birch but longer and more narrow. The bark is yellow or bronze and peels in narrow, curled strips. On very large trees, the bark may be reddish brown.

Yellow Birch often reaches great heights and is recognized by its thin trunk and rounded crown.

Yellow Birch bark typically peels into shaggy scrolls.

On the twigs of Yellow Birch are small, cone-like fruits.

When full grown, Sweet Birch bears a large, rounded crown with lower branches nearly parallel to the ground.

SWEET BIRCH

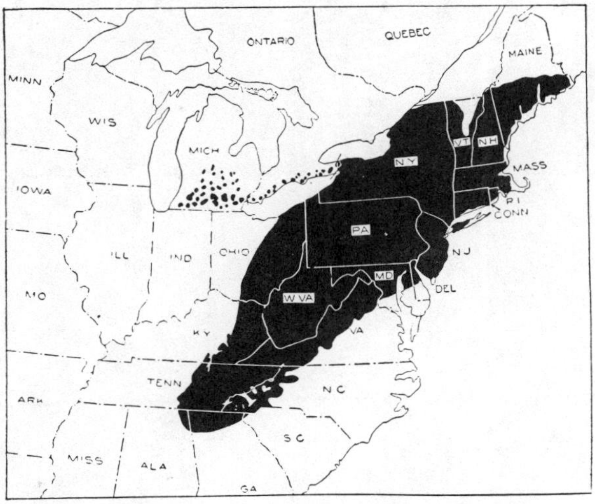

Sweet birch, when young, is black or reddish brown and does not peel. Bark appears in scaly plates on older trees.

River birch, another member of the species, is the only native birch in the southeastern United States.

Birch is a highly valued, excellent firewood. It is moderately easy to split and produces a high level of BTUs.

On mature Sweet Birch, the bark appears in scaly, irregular plates.

Sweet Birch leaves grow in alternate pairs off the sides of twigs and singly near the tips.

MAPLE (SUGAR)

SHAPE: Height reaches 100 feet with trunk diameter of 2-4 feet. In the forest it usually has a long, branchless trunk; in the open it will develop a large, rounded crown.

BARK: On young trees, bark is gray and smooth; on older trees, it is furrowed into irregular ridges or scales.

LEAVES: Leaves paired, heart shaped 3-5 inches in diameter, 3 or 5 lobed with the lobes long pointed and with blunt teeth. Dark green above, light green or pale and usually smooth beneath. Opposite.

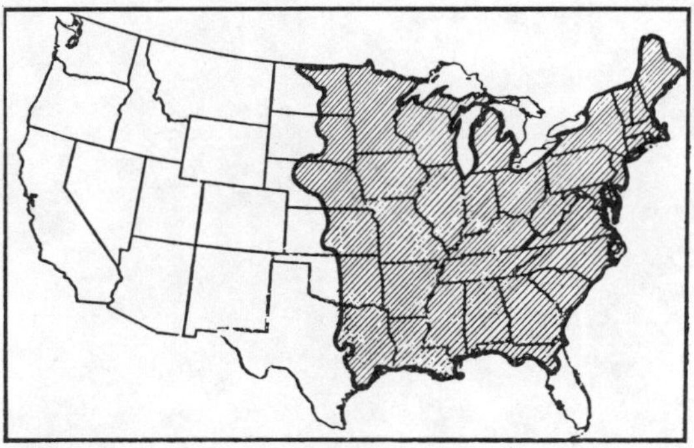

There are about 150 species of maples throughout the world, but only about a dozen are native to the United States. Among the most prominent and valued are the sugar, or hard, maples.

Hard maples are valued for furniture, flooring, handles, and other manufacturing. They also are a source of maple sugar and

Sugar Maple usually grows to a symmetrical and rounded shape.

syrup. In many locales they are planted extensively as shade trees.

The wood of sugar maples is very hard and straight grained and yields a high level of BTUs. It is excellent firewood.

The most common soft maples include silver maple, red maple, and striped maple.

Silver maple has grey, thin, smooth bark on large trunks. The leaves are slightly heart shaped 3-6 inches long, 5 lobed, long pointed and silvery white below, turning yellow in the fall.

Red maple has gray and smooth bark with heart shaped leaves 2-4 inches long, 3-5 lobes short pointed, dark green and shiny above, whitish and slightly hairy below, turning scarlet or yellow in the fall.

Striped maple has smooth, green bark with white stripes and leaves 5-6 inches long, 4-5 inches wide and turning yellow in the fall.

The bark of Sugar Maple is ash-gray in color.

Sugar Maple leaves are five-lobed and paired.

Two winged seeds and winter buds of Sugar Maple.

BOXELDER

Another member of the maple family is boxelder. It is a widely distributed tree in the United States with a gray or brown bark broken into narrow ridges.

Leaves are compound with usually 3 or 5, sometimes 7 or 9 oval or lance-oblong leaflets.

Boxelder is a fast growing but short lived tree, susceptible to decay and rot. The tree makes an inferior firewood.

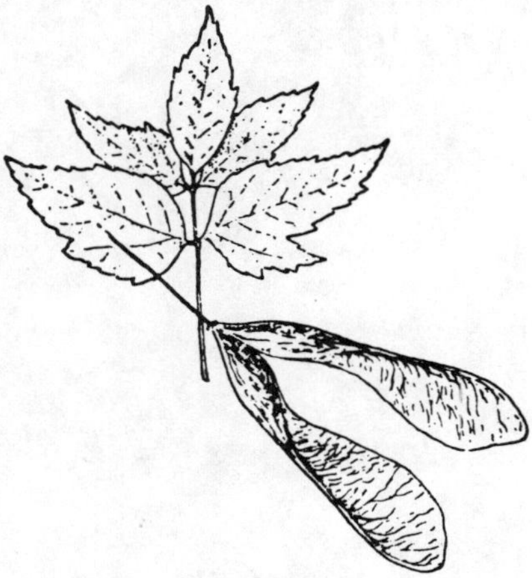

Boxelder leaves are compound with oval leaflets.

Black Cherry often develops a spreading, rounded crown with long, sometimes tortuous branches.

BLACK CHERRY

SHAPE: May reach 50-60 feet in height and 1-2 feet in diameter.

BARK: Smooth dark reddish to black bark on young trees; scaly and peels horizontally on older trees.

LEAVES: Narrow oval to oblong leaves 2-6 inches long and 1-2 inches wide with fine margin teeth. Alternate.

COMMON NAMES: Black cherry, wild cherry, rum cherry, mountain black cherry, cherry.

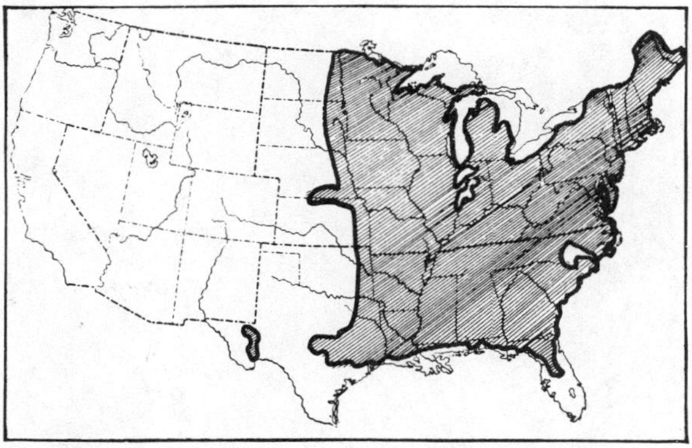

Black cherry grows under a wide range of climatic conditions, primarily in the eastern half of the United States, but also in scattered locations throughout the remainder of the United States and Canada.

Black cherry forms perfect flowers after the leaves are well developed. The fruit is a drupe about one-quarter the size of domestic cherries, with a bitter pulp and thick skin that is nearly black when ripe.

The sapwood of black cherry is narrow and varies in color from white to light reddish brown. Dark red gum stains are sometimes visible. The wood is naturally dull but takes on a fine luster when properly finished. The wood's color darkens with age, and it can be finished to bear a strong resemblance to genuine mahogany.

Black cherry is a straight grained, moderately hard and dense wood.

Black Cherry bark is dark with scaly and irregular plates.

FLOWERING DOGWOOD

SHAPE: A small, slow growing often crooked tree with a low, dense crown. Maximum size under favorable conditions is about 40 feet in height and 18 inches in diameter.

BARK: The bark is dark red brown to black, which on older trees breaks up into small quadrangular blocks.

LEAVES: 3-6 inches long and 1-2 inches wide, usually oval with a pointed end. Red in fall. Opposite.

COMMON NAMES: Cornel, boxwood.

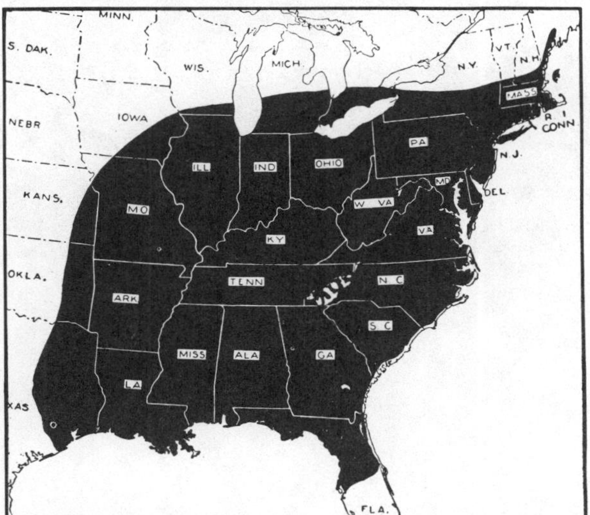

Flowering Dogwood grows a large, broad and irregular crown.

Flowering dogwood grows in most of the eastern United States. The tree has small, four part yellowish flowers surrounded by four snow white petals in the spring. In fall, leaves turn scarlet.

The sapwood of flowering dogwood varies in color from flesh colored, pinkish white to light pinkish brown. The heartwood is reddish brown to chocolate brown.

Dogwood is an extremely strong, hard, and heavy wood, but the cutting of live trees should be avoided because of their great beauty. Culling of dead or diseased dogwoods provides an outstanding firewood.

Broken into small quadrangular plates, the bark of Flowering Dogwood is dark red to black.

A showy spray of even flowers gives Flowering Dogwood its name.

Flowering Dogwood leaves are pointed and opposite.

BUCKEYE

SHAPE: Average trees reach 60-90 feet with a diameter of 2-3 feet and a rounded crown.

BARK: Grayish brown bark, furrowed and broken into large scales.

LEAVES: Compound, palm shaped with five elliptical leaflets each 4-6 inches long and 1-3 inches wide. Opposite.

COMMON NAMES: Yellow buckeye, Ohio buckeye, sweet buckeye, stinking buckeye, American horsechestnut.

Yellow buckeye is found in a very narrow range extending from southwestern Pennsylvania to southeastern Illinois.

Ohio buckeye has a greater range, extending into the southeastern and southwestern United States.

Yellow and Ohio buckeye are among the first trees to leaf out in the spring. Their flowers are yellow, or yellow white, 4-5 inches long. The principal difference between yellow and Ohio buckeye is in the size of the leaves which is smaller in Ohio buckeye and the fruit of Ohio buckeye is spiny while yellow buckeye is a smooth rounded, leathery capsule 2-3 inches long.

Both buckeyes produce large nuts that contain starch, but the nuts are not suitable as food because they contain a poison called aesculin.

The heartwood of buckeye is yellow or creamy white. The sapwood is white. The wood is moderately light and weak and is vulnerable to decay.

Yellow Buckeye bark is relatively smooth and even.

Ohio Buckeye bark is shallowly furrowed.

Buckeye leaves are narrow and opposite.

HOP HORNBEAM (IRONWOOD)

SHAPE: Tree or tall shrub, 10-50 feet in height, with a round topped crown.

BARK: Brown and scaly.

LEAVES: Elliptical, 2-4 inches long, 1-2 inches wide. Alternate.

The hornbeam is a small tree that grows throughout much of the United States. The leaves resemble those of elm, but the hornbeam is distinguished by its brown, scaly bark and hop-like fruits that hang in compact clusters. Its wood is very hard, heavy, strong, and makes an excellent firewood.

Hop Hornbeam although small, makes an excellent firewood.

BUTTERNUT

SHAPE: Generally a small to medium sized tree reaching 40-60 feet in height and 12-24 inches in diameter. Butternut growing in the open has a short trunk which divides into spreading branches close to the ground. Trees growing in a forest have straighter form, but still comparatively short trunks, short branches, and broad crowns.

BARK: On older trees, the bark is gray and furrowed.

LEAVES: Compound, usually about 15-30 inches in length, with 11-17 small leaflets attached to the stem. Opposite.

COMMON NAMES: Oilnut and white walnut.

Butternut is a member of the walnut family. Its range extends throughout the eastern half of the United States.

The fruit of the butternut is an oblong-ovoid nut with a very hard shell about two and a half inches long. The sweet oily kernel of the nut gives the tree its name.

The heartwood of butternut is light chestnut brown, sometimes tinged with red. The narrow sapwood is white to light brown and rarely more than one inch wide.

Butternut wood is light and fairly soft. It is a much lighter and inferior firewood to black walnut.

Straight boles of forest-grown Butternut.

The fruit of butternut is a sweet oily kernel covered with a hard shell.

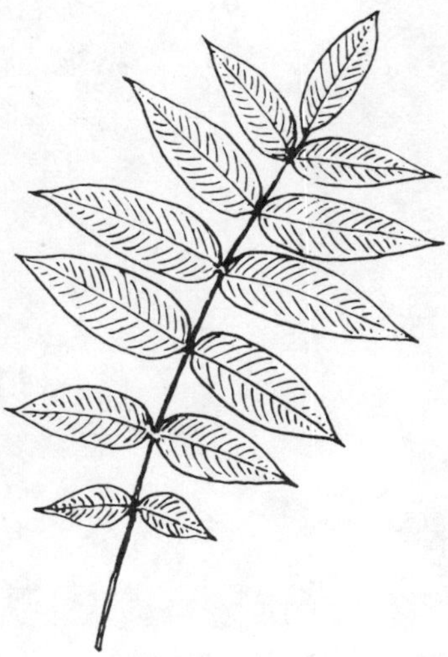

Butternut leaves are compound and opposite.

A large and full tree, the American Holly develops an extremely dense crown.

AMERICAN HOLLY

SHAPE: The size of holly varies considerably from bushy shrubs to small 40-50 feet trees on up to 70 feet in height and 2 feet in diameter. It usually has a short trunk and rounded crown.

BARK: Gray, often warty bark.

LEAVES: Green and leathery, 2-4 inches long by 1-2 inches wide, and varying greatly in shape. Alternate.

COMMON NAMES: Yule holly, Christmas holly and white holly.

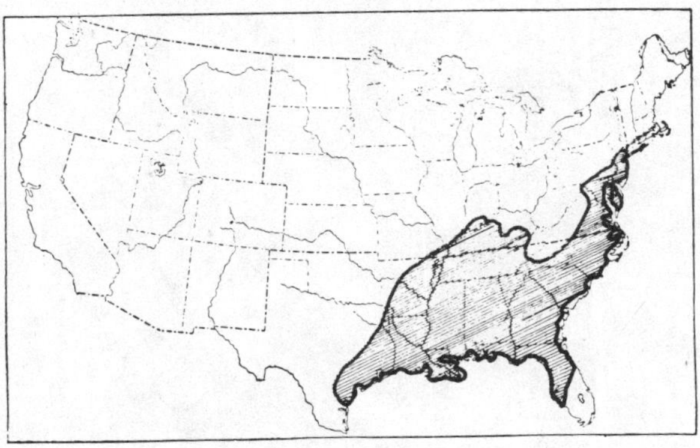

American holly is primarily found in the southeastern United States and into Texas and Oklahoma. Trees vary greatly in size, from bushy shrubs to small 40-50 feet trees and on up to 60-70 feet and two feet in diameter. Growth rate is poor. Holly may take 100-150 years to mature.

In southern bottomlands, American holly is usually found in small stream bottoms and on high ridges. In the uplands it is widely scattered on all but the driest and wettest sites. It grows best in Alabama, Arkansas, Louisiana and eastern Texas.

Flowers develop from April to June and are white or greenish. The fruit is a round drupe about one-half inch in diameter, maturing in November and December and is red or less often yellow and orange.

American holly has been described as the whitest wood known, having white sapwood and ivory white hardwood. The color of the heartwood is uniques among North American woods.

The wood of American holly is hard, moderately heavy, dense, and so it is a fine firewood.

American Holly bark is light gray and warty.

Often used for Christmas decorations, the American Holly berries are small and red.

The flowers of American Holly are small and white.

Tupelo develops a very dense, symmetrical crown.

TUPELO

SHAPE: Mature height may exceed 100 feet, with diameter of 40 or more inches.

BARK: Prominent ridges arranged inirregular block-like patterns, brown to black in color.

LEAVES: Oblong to elliptic form with a gradually narrowing base, 2-4 inches long and only one inch wide. The edges are usually smooth. Alternate.

COMMON NAMES: Blackgum, pepperidge, sour fum, gum tree, swamp blackgum, bay poplar, hazel pine, wild lime, gopher plum.

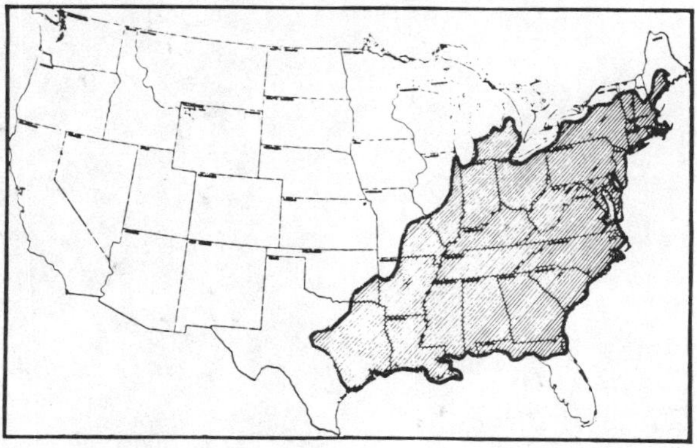

The species and varieties of tupelo in this country are black tupelo, swamp tupelo, water tupelo, and Ogeechee tupelo.

Black tupelo grows as a shrub or small tree in the dry eastern edge of the Great Plains but develops into a tree 100 or more feet in height in the lowlands of the South.

Swamp tupelo is limited to the Atlantic coastal plains and to estuaries from Maryland to Florida and along the Gulf of Mexico to southeastern Louisiana.

The range of water tupelo is similar to swamp tupelo, but the former also grows up the Mississippi Valley as far north as southern Illinois. It develops best in the bottom lands of major rivers and large streams.

The flowers of the most common species, black tupelo, appear in auxillary clusters in early spring and are greenish white. The fruit is a drupe 1/3 to 2/3 inch in diameter that matures in October. It is dark blue and has a bitter taste. When mature, the winter buds are about 1/4 inch long and are reddish brown.

Black tupelo is also called blackgum, pepperidge, sour gum, and gum tree. It is a moderately heavy wood with an interlocking grain. Color of the heartwood is brownish gray while the sapwood is lighter. Only a moderately strong and dense firewood.

Tupelo bark is deeply furrowed into long, thin plates.

The densely clustered leaves of Tupelo are oblong in form.

Yellow Poplar, or Tuliptree, is usually found mixed with other trees rather than in pure stands.

YELLOW POPLAR (TULIPTREE)

SHAPE: One of the largest eastern hardwoods, attaining heights of 150 feet with a diameter in trunk of 4-6 feet. Often has a straight trunk and pyramid shaped crown.

BARK: On young trees, the bark is dark green and smooth with a white, spotty appearance. Later, the bark breaks up into long, rough, interlacing, rounded ridges separated by deep gray furrows.

LEAVES: Leaves are shaped like a tulip, 4-6 inches long and 2 inches wide, bright green above and paler below. Alternate.

COMMON NAMES: Tuliptree, tulip poplar, white poplar, white wood, and poplar.

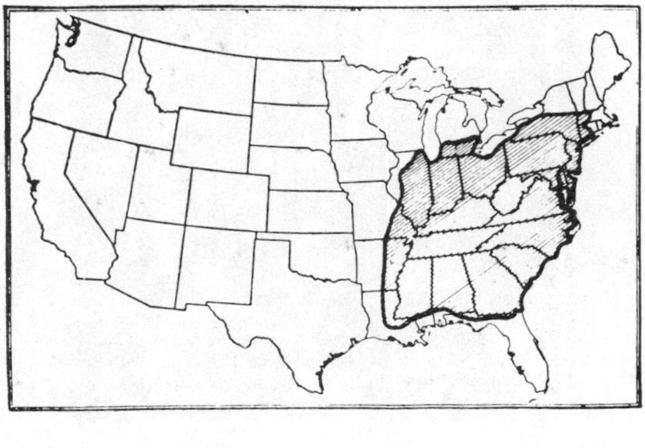

The mature yellow poplar has a striking appearance. In forests, its trunk is very straight and tall with no lateral branches for a considerable distance up the trunk. Though most yellow poplar forests have been cut within the last 50 years, some older trees occasionally can be found that have attained a height of 160 feet or more with a massive trunk eight feet or more in diameter. It probably is the tallest of all eastern broadleafed trees.

The heartwood of yellow poplar is tan or greenish brown, but may be quite variable with shades of purple, dark green, black, blue, and yellow. The sapwood is white and may have a striped appearance.

Yellow poplar is a soft, low density hardwood that does not rank with the better firewoods.

Yellow poplar bark is deeply furrowed when mature.

The flower of the Yellow Poplar resembles a tulip.

The Yellow Poplar's winter bud is dark red in color.

The American Sycamore is distinguished by its white, mottled bark.

AMERICAN SYCAMORE

SHAPE: Open grown trees are usually symmetrical and pyramidal during early years; later they develop large, irregular crowns with wide spreading branches. In the forest, trunks clear of limbs may extend to 70-80 feet, with fairly small crowns to 120 feet. Trunk diameters are usually 2-3 feet, but 10 deet has been recorded.

BARK: On young trees, bark is thin and smooth, becoming flaky at the base at about four years of age. Bark on older trees becomes whitish or brown peeling off in large flakes, sometimes in patches of green or gray.

LEAVES: Heart shaped, 4-8 inches long and wide, slightly 3 or 5 lobed, bright green and smooth above, paler and slightly hairy below.

COMMON NAMES: Buttonwood, buttonball, planetree.

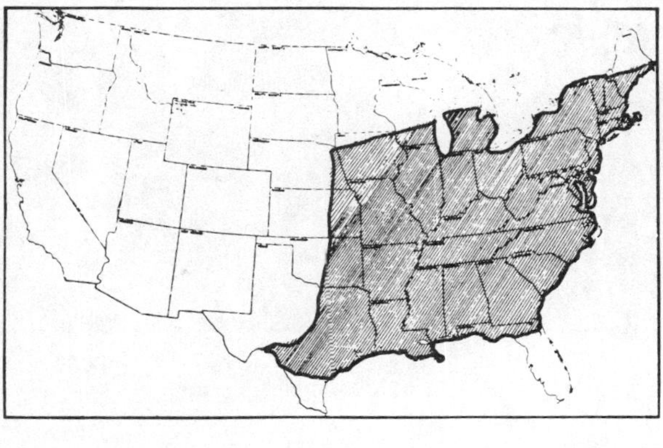

Sycamore grows to a large diameter than any other hardwood tree in America and is excelled in growth rate only by cottonwood, soft maple, and black willow.

It is not unusual for young sycamores on good sites to average one inch in diameter and six feet in height during each of the first ten years after planting.

The annual rings of sycamore are very distinct. The sapwood is not always distinguishable from the heartwood, but the sapwood generally is white to light yellow and the heartwood is light brown to dark brown or reddish brown.

The wood of sycamore is very difficult to split, but it is moderately strong, hard, and dense.

The inability of the bark to stretch over the expanding trunk causes the American Sycamore's patchy appearance.

American Sycamore leaves are often short and broad.

Hairy nut of American Sycamore.

Sweetgum develops a clean, straight trunk with dense foliage.

SWEETGUM

SHAPE: Mature trees are 1-3 feet in diameter and 80-120 feet tall. In the forest the trunk may be straight and take up two-thirds of the total height. Young trees usually have a compact crown but later it becomes open and round to flat on top with heavy branches.

BARK: Grayish brown with narrow, scaly ridges.

LEAVES: Leaves are star shaped, 5-7 inches wide with 5-7 pointed lobes and toothed margins, bright green turning red and gold in autumn. Alternate.

COMMON NAMES: Sweetgum, redgum, star-leaved gum, white gum, alligator tree, oppossum tree, and bilsted.

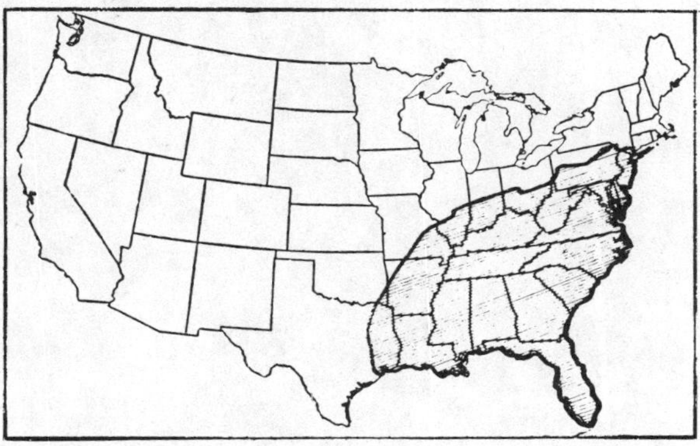

Sweetgum is a large tree of the eastern third of the United States, except on the northern border. It also grows in Mexico and central America south to Nicaragua.

Sweetgum usually grows in mixture with other hardwoods and pines. Growth varies greatly with conditions but may attain three feet per year in young trees.

Sweetgum produces small flowers in early spring. The fruit of the species is a spiky ball which lasts through much of the winter. It is composed of many small fruits, each with two long curved or spiny projections which open in autumn to release seeds.

The sapwood of sweetgum is wide and white to pinkish, sometimes with a bluish tinge produced by fungus infection. Heartwood varies through shades of red, reddish brown, and brown. The wood is moderately hard and heavy, but difficult to split.

The bark of Sweetgum is usually furrowed and gray.

OSAGE ORANGE

SHAPE: A large shrub or small tree averaging 30 feet in height, sometimes growing taller but rarely to 70 feet. It has a short trunk and a spreading crown.

BARK: Deeply furrowed orange-brown bark, with irregular, scaly ridges.

LEAVES: Ovate-lanceolate leaves 3-6 inches long, glossy dark green above and paler below. Leaves turn clear yellow in autumn. Alternate.

On one-year-old twigs, leaves are borne on short spurs which are keys to positive identification.

COMMON NAMES: Bodark, hedge, hedge-apple, osage-apple, horse-apple, mock-orange, bowwood, and yellow wood.

There is only one species of osage orange, including thorn and thornless varieties. The wood resembles black locust in appearance and physical properties. The tree has been planted extensively in hedges and windbreaks throughout much of the United States, except Minnesota, the Dakotas, and Montana where it is not winter-hardy.

The flowers of osage orange are four part, small, and green. They have no petals and appear after the leaves on the same spurs. After ripening the flowers become very fleshy, enlarged, and form a collective fruit which resembles an orange, yellow in color, 3-6 inches in diameter, and often weighing several pounds.

The sapwood is light yellow and narrow; the heartwood is golden yellow to bright orange, darkening sometimes with reddish streaks when exposed to air.

Osage orange ranks high in weight, strength, and hardness. The heartwood is the most decay-resistant of all North American timbers. It is an excellent firewood, despite its tendency to snap, crackle, and pop in the fireplace.

Typical furrowed bark of a mature Osage Orange.

Staminate flowers (male) in the axils of leaves on year-old branches of the Osage Orange.

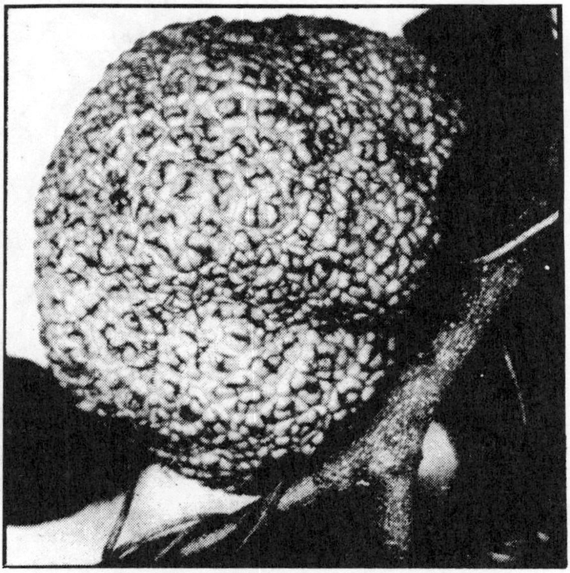

Ripe fruit of the Osage Orange often reaches six inches in diameter.

Common Persimmon can be found in the southern and eastern United States.

COMMON PERSIMMON

SHAPE: On better sites persimmon grows to heights of 30-60 feet or more, and to diameters of 12-24 inches.

BARK: On older trees, the bark is almost black, about one-quarter inch thick, and is broken into nearly square blocks.

LEAVES: Oblong to elliptic, 4-6 inches long by 2-3 inches wide, lustrous dark green above and paler below. Alternate.

COMMON NAMES: Chapote, eastern persimmon, Mexican persimmon.

Persimmon is a member of the ebony family of which some 300 species are recognized. It grows over all the southern states and through much of the remainder of the east and southwest.

The fruit of persimmon is a yellow, orange or brown globular, pulpy berry, 3/4 to 2 inches in diameter. The fruit is very astringent while green because of high tannin content, but it turns sweet and edible when thoroughly ripe.

The sapwood is creamy white when first cut but darkens to grayish or yellowish brown after exposure to air. The small heartwood is dark brown to jet black.

The principal and almost exclusive use for persimmon wood lies in the production of golf club heads. The wood is strong, hard, and heavy, and makes fine firewood.

Common Persimmon fruit is usually round and smooth.

Bark of the Common Persimmon is usually dark black and plated.

PECAN

SHAPE: Forest brown trees are usually medium sized and have straight trunks and medium to stout branches. Height may reach 100-140 feet with a trunk diameter of 2-4 feet.

LEAVES: The leaves of pecans are compound 12-20 inches long with 9-17 sessile, lanceolate to oblanceolate, opposite leaflets.

BARK: On young trees the bark is smooth and gray; on older trees it is grayish brown and usually broken into long, closely interlocked, narrow, rounded, scaly ridges separated by narrow furrows.

The pecan hickory group—pecan, water hickory, nutmeg hickory, and bitternut hickory—grown primarily in river bottoms of the lower Mississippi Valley and Gulf regions, although bitternut's range is most extensive, occuring in nearly all the eastern states.

The bark of young pecan hickories is smooth and gray; on old trees it is grayish brown and furrowed. The mature bark of water hickory is similar to pecan except it usually is browner with loose, shaggy, platy ridges. Mature bitternut bark is thin, pale, and has shallow furrows and low, narrow, interlocking ridges.

Leaves of pecans are 12-20 inches long with 9-17 leaflets. Fruit is 1-2 inches long and round to egg-shaped and covered by a hard husk. The meaty part of the nut is edible, and the pecan hickory nut is prized for its sweet taste.

The wood of pecan hickories is rated as strong, heavy, and very hard. It is somewhat lighter and denser than the true hickories, but stronger and denser than white oak, sugar maple, and white ash.

The slow burning, high BTU value, and hickory smoked flavor of the the pecan hickories have made them a popular fuel and charcoal producing wood.

Pecan bark is smooth and gray on young trees and furrowed in older ones.

Pecan leaves are long and narrow in opposite leaflets.

RED ALDER

SHAPE: Usually grows 40-60 feet, but mature trees may reach 120 feet with a diameter of two feet.

BARK: On young trees, bark is smooth and blue gray with white patches. Older trees are darker and bark lies in flat plates.

LEAVES: Leaves are ovate to elliptical, 2-6 inches long and 1-3 inches wide, dark green above and lighter underneath. Alternate.

The natural range of red alder extends along the Pacific Coast from southeastern Alaska to Santa Barbara, California.

Although it is found along streams in northern Idaho, usually it appears no further inland than 100 miles.

The flowers of red alder appear in the spring and grow larger as the leaves begin to appear. When the trees are about ten years old, seed production begins. The seed conelets are 1/2 to 1 inch long and 1/3 to 1/2 inch thick and contain 50-100 small nutlike seeds. New seedlings grow rapidly, reaching heights of as much as 18 inches the first year. Growth in following years can be even faster, but after ten years growth slows to about two feet per year.

Both sapwood and heartwood of red alder are nearly white when freshly cut. Afterwards, the wood changes to a reddish brown and when dry it may appear to be a light yellowish brown.

Red alder is moderately light, strong, and soft, but it is liked by many for fuel because it does not scatter sparks.

Red Alder leaves are ovate to elliptical in form.

Cones of the Red Alder contain small nut-like seeds.

MAGNOLIA (SWEETBAY)

SHAPE: 60-80 feet tall and 2-3 feet in diameter, with a pyramidal or round topped crown.

BARK: Brownish gray and scaly.

LEAVES: Leathery evergreen, oval to ovate, 5-10 inches wide. Alternate.

COMMON NAMES: Southern magnolia, evergreen magnolia, and bullbay. Sweetbay also is called swampbay, Southern sweetbay, evergreen magnolia, laurel magnolia, swamp magnolia, beaver tree, whitebay, and white laurel.

The magnolias are among the oldest tree species in the world. The range of southern magnolia includes a narrow strip about 100 miles wide along the coast of South Carolina, Georgia, Alabama, Mississippi, Florida, Louisiana, southern Arkansas, and east Texas. Sweetbay extends further north along the Atlantic coastal plain from Long Island to southern Florida. Unlike southern magnolia, the leaves of sweetbay are deciduous.

The flowers of magnolias are white, numerous, and very attractive. The fruit is an ovoid to cylindric cone, rose colored and 3-4 inches long by 1 1/2 to 2 inches wide.

The sapwood of magnolia is creamy white, and the heartwood is light to dark brown.

The wood of magnolia is moderately heavy, fairly hard, and straight grained.

Magnolia usually grows a round topped crown, pyramidal in form.

Magnolia bark is grayish and scaly.

TAMARACK (LARCH)

SHAPE: Generally grows to 100-175 feet in height and 2-4 feet in diameter. The trunks of mature trees are often free of limbs quite a distance up the tree; crowns are usually short, open and pyramidal. In summer, it is usually recognized by its light green foliage and in winter by the absence of needles.

BARK: On young trees the bark is thin and gray, becoming reddish brown to cinnamon brown and scaly on older trees. Bark on very old trees is flat plated and has deep irregular furrows.

LEAVES: The deciduous (falling off in autumn) needles are from 1-2 inches long, and three angled in cross section, and are a lustrous light green color. New needles generally appear in middle or late June with new short growth. Needles turn brown and drop off following fall frosts.

COMMON NAMES: Hackmatack larch, Montana larch, Western larch, mountain larch, Western tamarack, tamarack, larch.

Tamarack often attains heights of 175 feet with a straight and clean trunk.

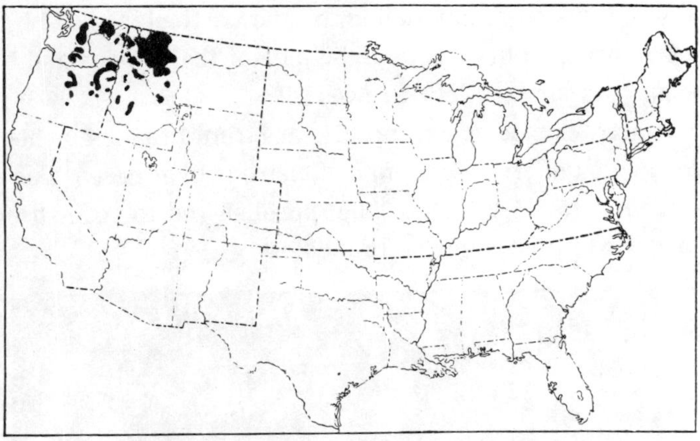

The tamarack is the only Northern member of the pine family to lose its leaves in winter — all other native species are evergreens.

The wood of the tamarack, unlike many of the other pines, is very heavy, hard, and durable, and makes an excellent firewood.

Tamarack has always been important in construction that requires a wood having high strength and hardness. It is used interchangeably with Douglas fir for general construction lumber or plywood. Knot free lumber is used extensively for interior paneling because of its good appearance and durability.

Tamarack has been used widely for railroad ties, for transmission poles, and for telephone poles especially where long lengths and high strengths are needed. Because of its hardness and resistance to splintering, there is some use of tamarack for mine shaft guides in deep underground mines. High speed underground mine hoists have run along these guides in lifting ore.

The sapwood of tamarack ranges from white to straw brown and is generally less than one inch wide. The heartwood is a distinct russet or reddish brown, sometimes resembling cedar.

The wood has a greasy feeling, but no particular odor or taste.

Tamarack splits easily and is one of the heavier and harder woods. It is an excellent firewood.

Tamarack's deciduous needles are from 1 to 1 3/4 inches long, three angled in cross section, and lustrous light green. The cones are from 1 to 1 1/2 inches long, purplish red to reddish brown, oblong, and stand erect on branchlets.

Tamarack bark is dull cinnamon in color with flat plates.

The needles of Tamarack are fine and flexible.

BALD CYPRESS

SHAPE: Large tree with swollen base growing to 150 feet and 8 feet in diameter, but often only 1-3 feet in diameter.

BARK: Reddish brown or gray with long scaly ridges.

LEAVES: Crowded and featherlike in two rows on slender horizontal twigs, flat 3/8 to 3/4 inch long, light yellow green or white beneath and shedding in fall (deciduous). Cones are 3/4 inch in diameter with hard scales. Alternate.

COMMON NAMES: Gulf cypress, red cypress, tidewater red cypress, white cypress, yellow cypress and cypress.

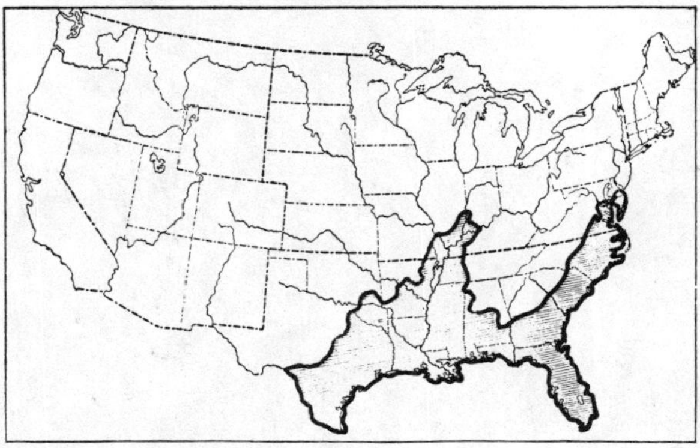

The heaviest stands of bald cypress occur in the swamps of the lower Mississippi Valley and Florida where pure stands

Bald Cypress is characterized by its massive trunk and irregular crown.

formerly covered large areas. It is a long lived tree that grows to large size throughout the southern United States. The tree develops an expanded conical base and an extensive root system.

Bald cypress produces some seed almost every year with seeds breaking away from the cone or the entire cone dropping to the ground. Cones contain pockets of red, sticky, liquid resin. Flower buds appear in winter with flowers, small and purple, growing in drooping clusters.

The fruit is a globose cone, 3/4 inch to one inch in diameter. Leaves are deciduous, alternate, and two ranked.

The sapwood of bald cypress is nearly white; heartwood ranges from light yellowish brown to dark brown, reddish brown, or chocolate.

The wood of bald cypress is of moderate weight, strength and hardness.

Bald Cypress bears woody and spherical cones containing resin.

The fluted trunk of the Bald Cypress grows best in southern swamps.

ENGELMANN SPRUCE

SHAPE: Mature trees are generally from 100-120 feet in height and from 18-36 inches in diameter.

BARK: Very thin, and broken into large purplish brown to russet red, loosely attached scales.

LEAVES: Needles are prickly and about one inch long, generally crowded on the upperside of branches, blue green in color.

COMMON NAMES: Columbian spruce, mountain spruce, silver spruce, and white spruce.

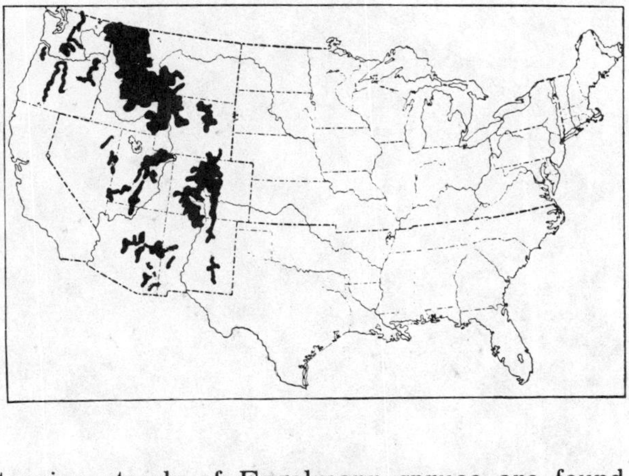

Extensive stands of Engelmann spruce are found at high elevations in the Rocky Mountains of nine western states and two Canadian provinces.

Towering to great heights, Engelmann Spruce is found at high elevations.

The heartwood of Engelmann spruce is nearly white, with an occasional slight tinge of red. The narrow sapwood is generally 3/4 to 2 inches wide in trees of saw timber size and is often difficult to tell from the heartwood.

Engelmann spruce is one of the lightest woods found in the United States, and is therefore an inferior firewood.

Engelmann Spruce cones are soft and flexible.

Engelmann Spruce bark is shaggy and loosely attached.

Sitka Spruce is one of the largest spruces and grows rapi lly.

SITKA SPRUCE

SHAPE: Vigorous fast growing tree usually occupying a dominant position in the forest. Tall, straight tree with an evenly tapering trunk to heights more than 200 feet and diameters of 4-16 feet, with an open conical crown.

BARK: Dark purplish and brown, thin, scaly and relatively smooth.

LEAVES: Needles are 5/8 to 1 inch long, sharp pointed, flattened and thin with two white bands on each side.

COMMON NAMES: Coast spruce, tideland spruce, and yellow spruce.

Sitka spruce has a 1,800 mile long range along the Pacific coast from northwestern California to Kodiak Island, Alaska. It is a fast growing tree, with cones usually 2-4 inches long, cylindrical, narrow, and light orange brown in color.

The sapwood of Sitka spruce is creamy white to light yellow and the heartwood is pinkish yellow to brown. The sapwood shades gradually into the heartwood; in mature trees it may be three to six inches wide and even wider in young trees. The wood has fine, uniform texture and a generally straight grain, no distinct taste or odor, and very few resin ducts. It is moderately hard and heavy.

BALSAM FIR

SHAPE: Balsam fir is one of the most symmetrical of the northestern coniferous species. It has a narrow pyramidal crown that ends in a tip like a spire.

BARK: On young trees the bark is thin with grayish patches and is smooth except for many resin patches. Older trees have bark that is reddish brown and has scaly plates.

LEAVES: Needles, usually 3/4 to 1 1/2 inches long, are flattened, shiny dark green above and silver below.

COMMON NAMES: Balsam, Canada balsam, eastern fir, and fir tree.

Balsam fir develops best in the cool climate of southeastern Canada and northeastern United States on deep sandy loam soils. It grows more slowly in the Lake States, probably because of lower rainfall.

The cones of balsam are 2-4 inches long, with short, round irregularly notched scales and pointed tips. Seeds are about 1/4 inch long, with broad purplish brown wings.

The sapwood of balsam fir is creamy white, and has little odor and taste.

The heartwood is gray to gray brown. The sapwood and heartwood are very low in resistance to decay, and balsam is one of the least durable woods. It is light in weight, density, and strength.

The bark of young trees is thin with grayish patches and is smooth except for many resin blisters. Older trees have bark that is reddish brown and scaly plates.

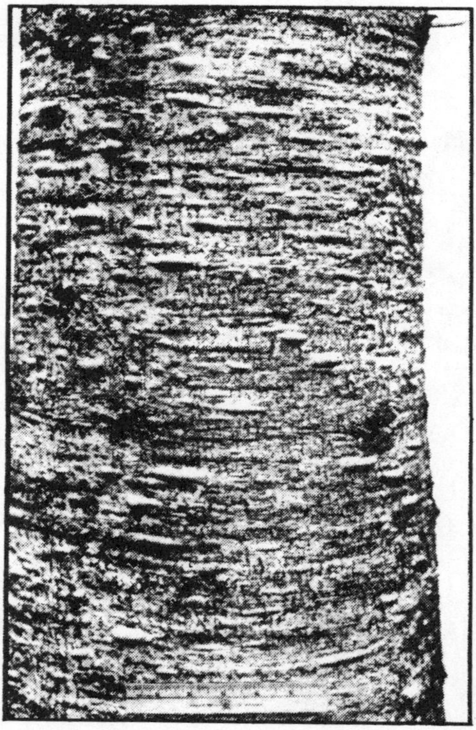

Balsam Fir bark is often marred with resin blisters.

Balsam Fir cones have irregular notched scales.

INCENSE-CEDAR

SHAPE: Reaches a height of 80-100 feet with diameters of 4-5 feet, with a spire like top.

BARK: Cinnamon colored and widely furrowed.

LEAVES: Twigs flattened, and wedge shaped 1/8 to 1/2 inch long, bright green with scale leaves.

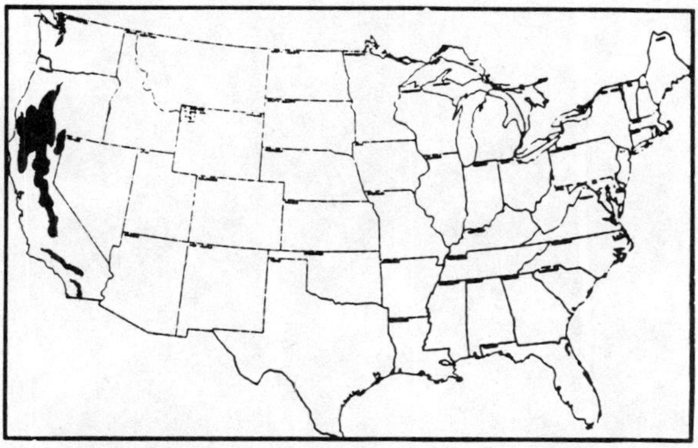

The range of incense-cedar extends from the southern slope of Mount Hood in Oregon southward to the mountains of Baja, California. The species grows in the coast range and west to the Sierra Nevada.

The sapwood of incense-cedar is white or cream colored. The heartwood is often light brown or light reddish brown, and is aromatic, spicy, smooth, and glistening. It is the heartwood with

Incensecedar is usually found in the coastal range of Oregon and California.

its outstanding resistance to decay which gives the wood of incense-cedar widespread use in a variety of lumber products.

Because of its outstanding durability and resistance to decay, incense-cedar is used extensively where humidity is high or moisture is ever present. Durability, and high insulation quality make the wood ideal for exterior siding where it is used extensively. These properties and its weathering ability adapt it well for log cabin siding.

Rich color, aromatic fragrance persuade many homeowners to use the wood for interior paneling and woodwork.

Incense-cedar also is used for closet lining. However, the wood is low in weight and hardness.

Bark of incense-cedar is reddish brown (cinnamon), thick and deeply furrowed into shreddy ridges.

The cones are 3/4 inch to one inch long and reddish brown.

Incensecedar bark is deeply furrowed and thick.

The light brown cones of Incensecedar hang from the ends of branches.

WESTERN WHITE PINE

SHAPE: Mature western white pines are generally the most impressive trees in the forest. They have tall, slightly tapering stems free of limbs up to as much as 100 feet, and symmetrical somewhat open crowns. Height may reach 200 feet with diameters of 7-8 feet.

BARK: On young trees bark is smooth and grayish green, turning dark or purplish gray on older trees, with alligator-like squares and prominent ridges.

LEAVES: Needles are blue green, about 3 inches long and grow in clusters of five. Color of the needles is a prime key to identification, since the blue green appearance is different from most other pines.

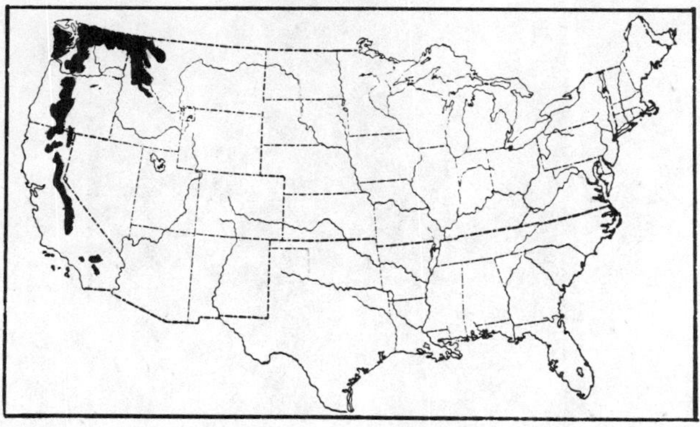

Western white pine grows on west coast mountain ranges. In its most northern range it grows from sea level to about 2500 feet elevation, but it attains greatest size in northern Idaho and

Western White Pine forms a symmetrical crown with short drooping branches.

adjacent sections of Montana, Washington, and British Columbia.

The sapwood of western white pine is nearly white, but sometimes pale yellowish white. Color of the heartwood ranges from cream to light brown or reddish brown and turns dark on exposure. The wood has the resinous odor typical of other pines, but not the characteristic taste.

The wood is generally straight grained, but soft and relatively light in weight.

Western White Pine bark is sharply broken into square blocks.

The needles of Western White Pine are pale and slender.

Western Redcedar develops a tapering trunk with drooping leaves.

WESTERN REDCEDAR

SHAPE: Attains heights of 200 feet and breast high diameters in excess of 8 feet (largest recorded diameter was 16 feet). The trunk in older trees rapidly tapers upward, and the crown is generally irregular but full with drooping branches which bend upward at the ends to present a distinctive look.

BARK: Generally about 3/4 inch thick, distinctive cinnamon red color and divided into thin interlacing ridges.

LEAVES: Very dark yellow green color, typical of arborvitae.

COMMON NAMES: Giant arborvitae, canoe cedar, arborvitae, shinglewood, Pacific redcedar, giant cedar, and cedar.

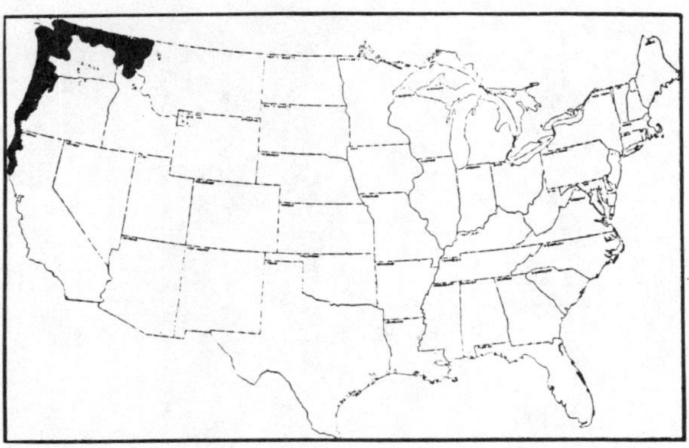

Western redcedar grows from the coast of southern Alaska to Northern California and east to the western slope of the Continental Divide. It has cones which are larger than the leaves and which become brown and woody when mature.

The wood has a distinct cedary odor and faint bitter taste and is almost always straight grained and easily split. The sapwood is nearly white and generally very narrow. The heartwood is reddish brown to dull brown. The wood is moderately soft and light.

Bark is cinnamon red and has thin interlacing ridges. Leaves are scalelike, 1/16 to 1/8 inch long, shiny and dark green. Cones are 1/2 inch long, pale brown and have leathery scales.

Western Redcedar leaves are flat and lacy.

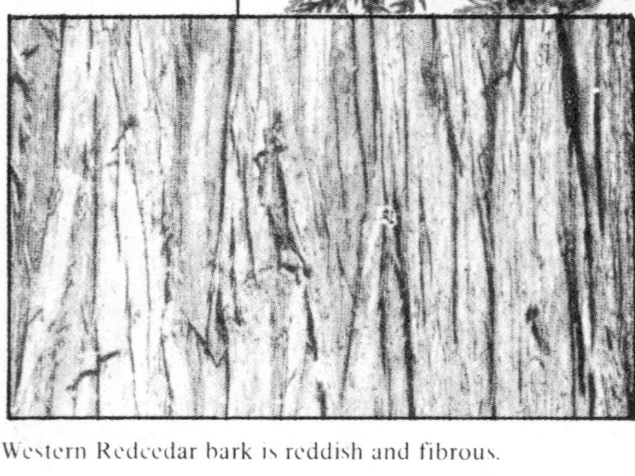

Western Redcedar bark is reddish and fibrous.

PONDEROSA PINE

SHAPE: Ponderosa is pyramidal when young, but the crown flattens at maturity. Mature trees may attain diameters in excess of 90 inches with heights of more than 200 feet.

BARK: On young trees, bark is furrowed and black, but turns yellowish brown to cinnamon and becomes heavily plated in older trees.

LEAVES: Needles are gray green to dark yellow green about 7 inches long and are almost always in clusters of three.

COMMON NAMES: Western pine, western yellow pine, western soft pine, western white pine, yellow pine, bull pine, and black jack pine.

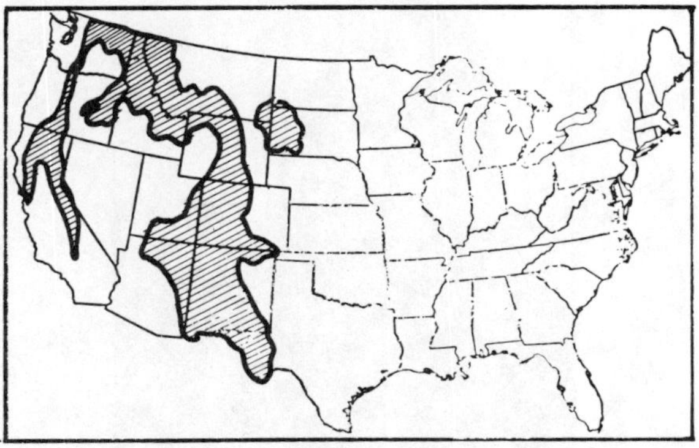

Ponderosa pine is the most widely distributed pine in North America, extending from British Columbia to Mexico and from

Ponderosa Pine forms a massive trunk with a spire-like head.

the Pacific Coast to Nebraska. As its wide distribution suggests, ponderosa pine grows on a large variety of soils. It grows best on well drained, deep sandy gravel and clay loams. It grows at elevations ranging from sea level to 9,000 feet, but is most commonly found at elevations between 4,000 and 8,000 feet. It grows generally in benches, plateaus, and west and south aspects.

Ponderosa pine is a long lived tree, sometimes reaching ages well beyond 300 years. One tree in eastern Oregon was more than 700 years old.

The sapwood of ponderosa pine is nearly white to pale yellow and generally two inches or more in width. The heartwood is yellowish to light reddish or orange brown and usually quite distinct from the sapwood.

The wood has a distinct resinous odor, but no characteristic taste. The grain is generally straight, but the wood is soft and light.

Ponderosa Pine bark is scaly and furrowed.

ALASKA CEDAR

SHAPE: Height ranges from 75-125 feet with diameters from 2-4 feet. Trunks may be swollen at the base and taper sharply.

BARK: On old trees, bark is thick, ashy brown or gray, and broken by shallow vertical furrows where the bark separates and curls into narrow flakes.

LEAVES: Small scalelike leaves are arranged in an overlapping pattern of four rows, giving a squarish cross section to the branchlets. Leaves are dull green or gray green and have sharp prickly, spreading points.

COMMON NAMES: Yellow cedar, Pacific Coast cypress, Alaska yellow cedar, yellow cypress, Nootka Sound cypress, Alaska ground cypress, Alaska cypress and canoe cedar.

The natural range of Alaska cedar extends from northern California through Oregon, Washington, and British Columbia, to Prince William Sound in Alaska.

Alaska cedar is confined to a cool, humid climate where growing seasons are short, but winter temperatures are not exceptionally severe. Snowfall is heavy over much of its range. Alaska cedar grows generally on poor sites. In Alaska, the species rarely grows on deep, well drained soils, apparently because it cannot compete with western hemlock and Sitka spruce on the better sites.

The wood of Alaska cedar resembles that of Port Orford cedar and Atlantic white cedar in appearance and properties. The heartwood is bright, clear, sulfur yellow. The sapwood is narrow, white to yellowish and is difficult to distinguish from the heartwood. The wood is fine textured and straight grained.

Alaska cedar is extremely durable. The heartwood is reputed to be resistant to attack by marine borers and so the wood has been used extensively in ship construction. The wood is moderately heavy and hard and is reasonably good firewood.

Alaska Cedar bark is ash brown or gray.

Leaves of Alaska Cedar are small and scalelike.

REDWOOD

SHAPE: Most mature redwoods are between 200-300 feet tall and range from 6-12 feet in diameter.

BARK: Dark, reddish brown and thick.

LEAVES: Needles on terminal shoots are small and scalelike; those on other branches are dark shiny green and about one half to one inch long.

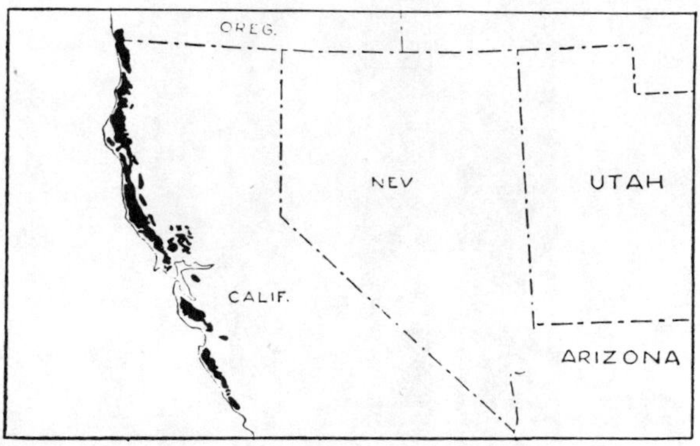

The bulk of old growth redwood goes directly into high value building construction materials. The wood is exceptionally straight grained and relatively free of oils and resins.

Redwood is a relatively light weight wood.

Known for its size, Redwood often towers to heights of 300 feet or more

Redwood bark is reddish and thick.

Leaves of the lower branches of Redwood stand out stiffly from the twigs.

PORT ORFORD CEDAR

SHAPE: Mature trees are 4-6 feet in diameter and up to 200 feet tall.

BARK: 6-10 inches thick in old trees, silver brown, and divided into rounded ridges separated by deep, irregular furrows.

LEAVES: Scalelike, small, paired, and blunt.

The natural range of Port Orford cedar is restricted to the Pacific coast from Coos Bay, Oregon, to Eureka, California. The best stands are not more than 10-40 miles inland.

The sapwood of Port Orford cedar is nearly white to a pale yellow white (often not clearly distinguishable from the heartwood) and 1-3 inches wide. The heartwood is yellowish white to pale yellowish brown.

The wood of Port Orford cedar is generally straight grained and moderately light in weight and hardness.

Port Orford Cedar bark is thick and silver brown.

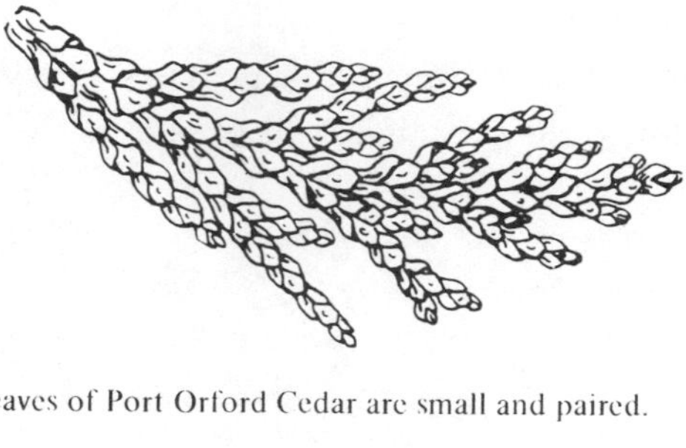

Leaves of Port Orford Cedar are small and paired.

JACK PINE

SHAPE: Mature trees are 55-65 feet tall and 8-10 inches in diameter at breast height.

BARK: Thin, brown, slightly tinged with red or dark gray, and irregularly divided into scaly ridges.

LEAVES: Needles grow in pairs 3/4 to 1 1/2 inches long. They are yellow green, flat or slightly concave on the inner surface, and stout.

COMMON NAMES: Gray pine, scrub pine, and banksiana pine.

The natural range of jack pine extends from the Maritime Provinces west to Alberta and the District of MacKenzie and southward to the Lake States of Minnesota, Wisconsin, and Michigan. Local stands are found also in many other midwestern and eastern states.

Jack pine cones are 1 1/2 to 2 inches long, oblong and light brown, usually pointing forward and often strongly incurved with scales well developed only on the outer face. Cones may remain tightly closed without releasing seeds for as many as 25 years or more in the northern part of its range.

Jack pine generally grows in areas with warm to cool summers, very cold winters, rather low rainfall, light sandy soils and rolling topography. In the northern Lake States and parts of central and western Canada, jack pine stands cover large areas. The species generally develops a wide spreading and only moderately deep root system. However, a taproot frequently develops which may penetrate below 9 feet.

In well stocked stands, jack pine develops into a rather slender tree with a narrow open crown. As the stand closes and matures, up to two-thirds of the lower crown may die. Open grown jack pines have poor form and the crowns assume a spreading appearance with live branches persisting almost to the ground.

The species is relatively windfirm although wind breakage may be common in ice storms. Early spring fires readily kill many young trees, although mature jack pines are relatively fire resistant.

The sapwood of jack pine is nearly white and generally makes up at least 50 per cent of the volume. The color of heartwood varies from light brown to a distinct orange cast. Jack pine lumber is generally knotty, has a coarser textured, more resinous appearance than white or red pine, and is used principally for pulpwood. The wood is moderately light in weight and hardness.

Jack Pine bark is furrowed into scaly ridges. The cones are oblong and light brown, they point forward from the twig.

EASTERN WHITE PINE

SHAPE: Reaches heights of 100 feet or more and diameters of 3-6 feet. It has a sturdy, gradually tapering trunk and horizontal limbs.

BARK: On young trees, bark is thin, smooth, and greenish brown. Later, the bark darkens and becomes thicker and deeply furrowed.

LEAVES: Distinctive bluish green needles in bundles of five, 3-5 inches long that remain on the tree from 3-5 years.

COMMON NAMES: White pine, northern white pine, northern pine, soft pine, spruce pine.

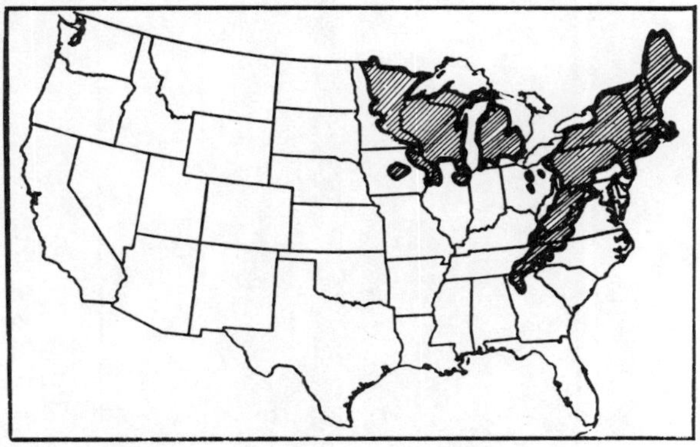

Eastern White Pine is tall and slender.

The natural range of eastern white pine extends from Newfoundland west to south eastern Manitoba, southward through the Lake States to Pennsylvania, New York, and New England. It also extends southward to Georgia.

White pine is the fastest growing northeastern forest tree. It grows taller than any pine in the east except for the sugar pine. It can be separated from all other eastern pines by its bluish green needles in bundles of five, 3-5 inches long. The cylindrical cones, 4-8 inches long, have thin usually gummy scales containing small winged seeds that are released at the end of the second season's growth.

The original stand of eastern white pine in the United States and Canada was estimated at 750 billion board feet — more than the current estimated volume of all sawtimber in the United States east of the Rockies. Heavy cutting in the Lake States and in the east greatly reduced the stands of white pine, but replanting during the past several decades has brought the total back to a higher level.

The heartwood of eastern white pine is light brown or reddish brown and turns darker on exposure. The sapwood, narrow to medium in width depending upon the age and vigor of the tree, is white to pale yellowish white. The wood is usually straight and evenly grained, light in weight, and moderately soft.

Bark on Eastern White Pine is cut with vertical ridges.

Eastern White Pine cones are long and tapered.

EASTERN HEMLOCK

SHAPE: Often reaches a height of 80-100 feet with a diameter of 2-3 feet. Mature trees usually have straight, tapering trunks with dense long crowns, and crowns of old trees tend to be ragged.

BARK: On young trees, bark is flaky and becomes crossed by deep furrows as the tree matures. On old trees, the bark is often 2-3 inches thick at the stump, deeply furrowed, and reddish brown.

LEAVES: Needles are short, flat, and narrow. They are about 1/5 inch wide and 1/3 to 2/3 of an inch long. New needles are light green but soon change to a dark lustrous green on the upper surface and whitish green underneath.

COMMON NAMES: Hemlock, hemlock spruce, spruce pine, and Canadian hemlock.

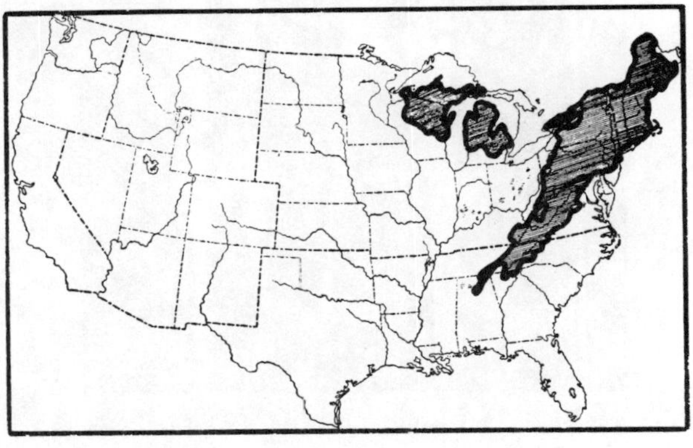

Rugged and massive, Eastern Hemlock has a clean and straight trunk.

The natural range of eastern hemlock extends from Nova Scotia to Minnesota and from New England through the Middle Atlantic states to Indiana.

Because hemlock is tolerant of shade, the lower branches remain alive for many years. When the branches die they persist for many years leaving hard, flint like knots. Because of this poor self pruning ability even in dense stands, clear trunks more than 30 feet long are uncommon except in very old trees.

The cones of hemlock are among the smallest of the pine family, 1/2 to 3/4 inches long. They are pale green until maturity in the autumn when they become dark brown.

Eastern hemlock wood is light brown to buff. The sapwood is usually not much different from the heartwood, but sometimes the last few annual rings near the bark are lighter in color than the inner rings. Freshly cut hemlock has a sour odor but when seasoned the wood is odorless. It is moderately light and weak.

Eastern Hemlock bark is thick and furrowed.

Needles of Eastern Hemlock are narrow and flat.

EASTERN SPRUCE

Three spruces — white, black, and red — are known as Eastern spruces. The natural range of white spruce extends from Newfoundland and Labrador west across Canada to northwestern Alaska and southward through New England, New York, Michigan, Wisconsin, and Minnesota.

Black spruce is one of the most abundant conifers in North America and has a natural range similar to that of white spruce. Black spruce is primarily a tree of the northern interior swamps and lowlands. It is commonly found in cold, poorly drained bogs throughout its range and under such conditions forms dense slow growing pure stands.

Red spruce extends from Nova Scotia and New Brunswick southward through New England and eastern New York. It continues southward at high elevations in the Appalachians to western North Carolina and eastern Tennessee.

White Spruce bark is grayish brown and thin.

White Spruce needles are spiky and blue green.

WHITE SPRUCE

SHAPE: May attain a height of 120 feet with diameters of 48 inches.

BARK: Gray or brown, thin and scaly.

LEAVES: The needles are 1/3 to 3/4 inches long, four sided, blue green, and sharp to the touch.

White spruce cones are usually longer than those of red or black spruce, ranging from 1 1/2 to 2 inches. They are light brown with thin flexible scales and usually flat at the end.

Bark of white spruce is gray or brown and thin and scaly.

Bark of black spruce is grayish brown and thin and scaly.

Needles of black spruce are 4 angled, 1/4 to 5/8 inch long and pale, blue green. Cones are 3/4 to 1 1/2 inches long, dull gray brown with scales. Bark of red spruce is reddish brown. Needles are 4 angled, 1/2 inch long, dark green and shiny.

Cones are 1/4 to 1 1/2 inches long, light reddish brown and shiny.

Red Spruce grows an open and wide-spreading crown.

RED SPRUCE

Bark of red spruce is reddish brown. Needles are four angled, ½ inch long, dark green and shiny.

Cones are ¼ to 1½ inches long, light reddish brown and shiny.

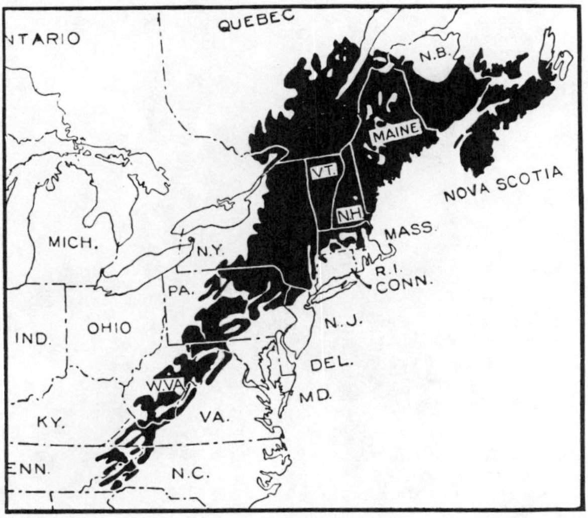

Red Spruce bark is hard and firm and dark brown in color.

Red Spruce cones are glossy and reddish brown.

BLACK SPRUCE

Needles of black spruce are four angled, 1/4 to 5/8 inches long and a pale, blue green.

Cones are 3/4 to 1½ inches long, dull gray brown with scales.

Black Spruce bark is dark and scaly.

Black Spruce bears pale needles and dark brown cones.

APPENDIX

Wood is often sold by the cord or a fraction of a cord, so it pays to know the difference between a *standard cord,* a *face* cord, a *long* and a *short* cord, a *solid* cord, and perhaps others commonly used by wood sellers.

Always specify a standard cord when you order your wood from a dealer. Make sure that what you get is indeed a pile of wood that is four feet high, eight feet wide and four feet long.

Measure the wood pile and make certain that it is tightly stacked and that all of the logs lie in the same direction, otherwise you may be paying for air space instead of wood. Remember, even when wood is carefully piled, it contains only 80 cubic feet of solid wood, although its dimensions may total 128 cubic feet. The size and straightness of a wood in a pile determines how much wood you get. Crooked pieces add air to the space, while large bulky pieces add up to more wood even though the spaces may appear larger.

Generally, pound for pound, most kinds of wood produce just about the same amount of heat, but some woods are lighter than others. A cord of aspen weighs about half as much as a cord of dry white oak and therefore has about one half the heat value.

SPECIES	Coaling Qualities	Heavy Smoke	Splitting Ease	Starting Ease	Sparks
Apple	*Great*	*No*	*Good*	*Poor*	*Few*
Ash	*Good*	*No*	*Good*	*Fair*	*Few*
Beach	*Good*	*No*	*Poor*	*Poor*	*Few*
Birch (wh)	*Good*	*No*	*Good*	*Good*	*Some*
Cherry	*Great*	*No*	*Good*	*Poor*	*Few*
Cedar	*Poor*	*Yes*	*Good*	*Great*	*Many*
Elm	*Good*	*Some*	*Poor*	*Fair*	*Very Few*
Hemlock	*Poor*	*Some*	*Good*	*Good*	*Many*
Hickory	*Great*	*No*	*Good*	*Fair*	*Some*
Locust b.	*Great*	*No*	*Poor*	*Poor*	*Very Few*
Maple	*Great*	*No*	*Good*	*Poor*	*Few*
Oak r.	*Great*	*No*	*Good*	*Poor*	*Few*
Pine w.	*Poor*	*Some*	*Good*	*Great*	*Some*
Spruce	*Poor*	*Yes*	*Poor*	*Good*	*Some*
Willow	*Poor*	*No*	*Good*	*Fair*	*Few*

INDEX

Alaska cedar. 169-170
Alder 135-136
Alligator bark 25
Alligator tree 124-126
Alternate 20
American basswood 51-54
American beech 59-63
American elm 49-50
American holly 108-111
American horsechestnut . 101-102
American sycamore 120-123
Ash. 43-46
Aspen 79-80

BTUs 9, 13, 17
Bald cypress 143-146
Balsam fir 153-154
Bark (key chart) 27
Barks 21-27
Basswood. 51-54
Beaver tree. 137-138
Beech 59-63
Bigtooth aspen 79-80
Bilsted. 124-126
Birch. 81-82
Bitternut hickory 133-134
Black cherry. 94-96
Black cottonwood 71-74
Black locust 64-67
Black spruce. 187-193
Black tupelo. 112-115
Black walnut 55-58
Black willow. 69-70
Black jack pine 165-167
Bodark 127-129
Bow wood 127-129
Boxelder 93
Boxwood. 97-100
Buckeye 101-102
Bullbay 137-138
Butternut. 105-107
Buttonball 120-123
Buttonwood. 120-123

Canoe birch 81-82
Canoe cedar . . .162-164, 169-170
Cedar . . . 155-158, 162-164, 169
Chapote. 130-132
Cherry. 94-96
Christmas holly. 108-111
Columbian spruce 147-149
Common persimmon . . . 130-132
Cord 9, 195
Cornel. 97-100
Cottonwood. 71-74
Cypress 143-146

Deciduous trees. 19
Dogwood. 97-100

Eastern cottonwood. 71-74
Eastern fir 153-154
Eastern hemlock 183-186
Eastern persimmon 130-132
Eastern poplar 71-74
Eastern spruce 187-188
Eastern white pine. 179-182
Elm 47-48

Englemann spruce 147-149
Evergreen magnolia 137-138
Evergreens 19

Face cord. 9, 195
Fir 153-154
Fire, starting. 18, 196
Firewood, heating
value chart 14
Firewood, moisture 16-17
Firewood, seasoning. 17
Firewood, sources 11-13
Firewood, splitting . . 16-17, 196
Firewood, stacking 17, 195
Firewood, types 13
Firewood, weight . . . 13-15, 195
Fissured. 22-23
Flowering dogwood 97-100
Furrowed. 22, 23

Giant arborvitae 162-164
Gulf cypress 143-146

Hackberry 75-77
Hackmatack larch 139-142
Hard maple 89-92
Hardwoods. 13-15
Heartwood. 21
Hedge-apple 127-129
Hemlock 183-186
Hickory. 37-42
Holly 108-111
Hop hornbeam 103-104
Hornbeam 103-104
Horse-apple 127-129

Incense-cedar 155-158
Ironwood. 103-104

Jack pine 177-178

Larch 139-142
Largetooth aspen. 79-80
Laurel magnolia 137-138
Leaf scars. 21
Leaves. 19-21
Linden 51-54
Locust. 64-67

Magnolia 137-138
Maple 89-92
Mock orange. 127-129
Mockernut hickory 37-38
Mountain black cherry 94-96
Mountain larch 139-142
Mountain spruce 147-149

Nutmeg hickory 133-134

Oaks 29-36
Ogeechee tupelo 112-115
Ohio buckeye 101-102
Oilnut 105-107
Opossum tree 124-126
Opposite 20
Osage-apple 127-129
Osage orange 127-129

Pacific redcedar. 162-164
Paper birch. 81-82
Pecan 133-134
Persimmon. 130-132
Pignut hickory 37-38
Pine . 159-161, 165-167, 177-178
Planetree 120-123
Ponderosa pine 165-167
Poplar 116-119
Popple. 79-80
Port Orford cedar 175-176

Quaking aspen 79-80

Red alder. 135-136
Red cypress 143-146

Red maple 89-92
Red oak. 30, 34, 35-36
Red spruce. 187-192
Redgum. 124-126
Redwood. 170-173
Resin 21
Ridged 22-23
River birch. 87

Sapwood 21
Shagbark hickory . . 37-38, 39-42
Shellbark hickory 37-38
Silver maple 89-92
Silver spruce. 147-149
Sitka spruce 150-152
Softwoods 13-15
Southern magnolia. 137-138
Spruce. 147-149, 150-152
Striped maple. 89-92
Sugar maple 89-92
Sugarberry 75-77
Swamp tupelo. 112-115
Swampbay 137-138
Sweet birch 86-88
Sweet buckeye 101-102
Sweetbay. 137-138
Sweetgum 124-126
Sycamore. 120-123

Tamarack. 139-142
Tuliptree 116-119
Tupelo 112-115

Walnut 55-58
Water hickory. 133-134
Water tupelo. 112-115
Western larch 139-142
Western redcedar. 162-164
Western tamarack 139-142
Western white pine 159-161
White ash. 43-46
White cypress 143-146
White gum 124-126
White holly 108-111
White laurel 137-138
White oak 30, 31-33
White poplar. 116-119
White spruce. . .147-149, 187-189
White walnut 105-107
White wood 116-119
Whitebay 137-138
Willow. 69-70

Yellow birch. 83-85
Yellow buckeye 101-102
Yellow cypress .143-146, 169-170
Yellow poplar. 116-119
Yule holly 108-111